Newton Gresham Library
Sam Houston State University
Huntsville, Texas 77341

WITHDRAWN

WITHDRAWN

REF.

AF697669

GEOSTATISTICAL GLOSSARY AND MULTILINGUAL DICTIONARY

INTERNATIONAL ASSOCIATION FOR MATHEMATICAL GEOLOGY
STUDIES IN MATHEMATICAL GEOLOGY

1. William B. Size (ed.): *Use and Abuse of Statistical Methods in the Earth Sciences*

2. Lawrence J. Drew: *Oil and Gas Forecasting: Reflections of a Petroleum Geologist*

3. Ricardo A. Olea (ed.): *Geostatistical Glossary and Multilingual Dictionary*

GEOSTATISTICAL GLOSSARY AND MULTILINGUAL DICTIONARY

Members of the 1984-1989 IAMG Committee on Geostatistics

RICARDO A. OLEA, editor

New York Oxford OXFORD UNIVERSITY PRESS 1991

02118007541244

Oxford University Press

Oxford New York Toronto
Delhi Bombay Calcutta Madras Karachi
Petaling Jaya Singapore Hong Kong Tokyo
Nairobi Dar es Salaam Cape Town
Melbourne Auckland

and associated companies in
Berlin Ibadan

Copyright © 1991 by Oxford University Press, Inc.

Published by Oxford University Press, Inc.,
200 Madison Avenue, New York, New York 10016

Oxford is a registered trademark of Oxford University Press

All rights reserved. No part of this publication may be reproduced,
stored in a retrieval system, or transmitted, in any form or by any means,
electronic, mechanical, photocopying, recording, or otherwise,
without the prior permission of Oxford University Press.

Library of Congress Cataloging-in-Publication Data
Geostatistical glossary and multilingual dictionary / edited by Ricardo A. Olea
(the 1984-1989 IAMG Committee on Geostatistics).
p. cm. — (Studies in mathematical geology ; 3)
Includes bibliographical references (p.).
ISBN 0-19-506689-8
1. Geology—Mathematics—Dictionaries—Polyglot.
2. Geology—Statistical methods—Dictionaries—Polyglot.
3. Dictionaries, Polyglot.
I. Olea, R. A. (Ricardo A.)
II. International Association for Mathematical Geology.
Committee on Geostatistics. III. Series.
QE33.2.M3G46 1991 550′.72—dc20 90-7409

9 8 7 6 5 4 3 2 1

Printed in the United States of America
on acid-free paper.

Members of the 1984–1989 Committee on Geostatistics
of the International Association for Mathematical Geology
who participated in the project

GEORGE CHRISTAKOS
Harvard University
Cambridge, Massachusetts, U.S.A.
presently at
University of North Carolina
Chapel Hill, North Carolina, U.S.A.

ISOBEL CLARK
Geostokos Ltd.
London, England

MICHEL DAVID
Geostat Systems International Inc.
Montréal, Québec, Canada

ANDRÉ G. JOURNEL
Stanford University
Stanford, California, U.S.A.

DANIEL G. KRIGE
University of Witwatersrand
Johannesburg, South Africa

RICARDO A. OLEA (Chairman)
Kansas Geological Survey
Lawrence, Kansas, U.S.A.

FOREWORD TO THE SERIES

This series of studies in mathematical geology provides contributions from the geomathematical community on topics of special interest in the earth sciences. As far as possible, each volume in the series will be self-contained and will deal with a specific technique of analysis. For the most part, the results of research will be emphasized. An important part of these studies will be an evaluation of the adequacy and the appropriateness of present geomathematical and geostatistical applications. It is hoped the volumes in this series will become valuable working and research tools in all facets of geology. Each volume will be issued under the auspices of the International Association for Mathematical Geology.

U.S. Geological Survey
Reston, Virginia

Richard B. McCammon

PREFACE

Throughout the preparation of this glossary, beginning with the arduous task of agreeing on a definition for geostatistics itself, our aim has been to present a useful reference for students, scientists, engineers, and others concerned with the meaning of the terms found in the geostatistical literature and heard in professional practice. Definitions and entries, which differ in levels of rigor and complexity, have been prepared in order to provide help to a readership that may range from inexperienced practitioners to advanced researchers.

The eclectic nature of this undertaking inevitably has produced problems in choosing the entries and deciding how much information to include. We have not defined words like *location* that are used in geostatistics with the same meaning as given in a standard dictionary. Instead, we have focused on the technical meaning of words such as *geometrical anisotropy* or *kriging*, which appear in few if any dictionaries with the meanings they are given in geostatistics. A self-contained, parsimonious, and easy to read geostatistical glossary, however, could not be produced without including several technical terms, such as *heteroscedastic* or *Lagrange multiplier*, that are taken from other areas like calculus, probability, and statistics. Definitions of these basic, nonexclusively geostatistical entries may be unnecessary for the mathematically sophisticated reader, but useful for those with a more limited background. We have concentrated on the mathematical aspect of geostatistics, but have not lost sight of the applied side; however, we have not included those terms that are used in the particular subject areas where geostatistics is applied. For example, the mathematical definition of terms that a forester might encounter when using geostatistics to solve a problem of tree sampling are included, but the technical vocabulary peculiar to the science forestry itself is not. In some instances, a particular entry may be missing simply because we did not happen to think about it.

The committee did not invent the terms included here, but merely collected the words and their meaning as encountered in actual usage, regardless of propriety. Some of the terms and definitions have been considered controversial even by some of the committee members who decided to contribute to this volume. The inclusion of a particular meaning in this glossary does not imply that this definition is in some sense official, or that it is unanimously endorsed by the geostatistical community. When terms have been given more than one acceptable meaning, we have listed them all, if possible in descending order of generality, the length of time a particular meaning has been used, or the frequency of usage. Numerous references were consulted to assure as much conformity of usage as possible.

The emphasis in the glossary section is on definitions of terms rather than on pronunciation, etymology, or syllabication. All headings are printed in boldface capital letters starting at the left margin, and are alphabetized on a Roman letter basis. Spaces, hyphens, commas, and apostrophes have been ignored for the purpose of establishing the alphabetical sequences of words. Except for cross-references, the definitions of terms are printed in lightface lower-case text. Key words that occur in a definition and which are defined elsewhere in the glossary are printed in boldface type to help the reader if the definition of a word leads to the search of another. The definitions of few terms have been supplemented with illustrations or simple numerical examples.

Geostatistics is a multidisciplinary field that has generated its own notation that also has borrowed systems of notation from other fields. These borrowed notations sometimes tend to be inconsistent. In addition, geostatistical notation is not yet standardized. In this glossary we have attempted the impossible task of reconciling the desire for consistency with the need to preserve traditional usages. Each mathematical symbol that may be required to explain a term is defined in each entry, which it is hoped should resolve any remaining inconsistencies.

Because the International Association for Mathematical Geology (IAMG) is a worldwide organization, we have also prepared a multilingual dictionary section. This section may at least partially overcome the inconvenience that we have caused in the first section for some users for whom English is not their first language. Seven languages were selected for the multilingual dictionary in addition to English: Chinese, French, German, Greek, Portuguese, Russian, and Spanish. These eight languages include the four most-spoken tongues in the world, and account for native languages of more than 90 percent of the participants at the last International Geostatistics Congress. While most of the terms in the glossary were originally formulated either in French or English, they have been translated into most of the tongues included here throughout the years. In preparing the dictionaries, the committee gathered information from published sources with the same open mind that guided the preparation of the glossary. If more than one translation is widely used, we list all synonyms rather than favoring a specific usage. The committee accepted translations especially coined for this glossary only when no evidence of a translation could be found in the literature. In general such terms are limited to the most recent methodologies, whose terminology has not yet found its way into the scientific literature except in the tongue in which they were originally presented. These new translations were prepared for the sake of completeness and should in no way be considered as the definitive terms. The committee is willing to see some of its translations replaced by better equivalences and others supplemented with alternative synonyms.

The dictionary, arranged alphabetically by English terms, is intended to help readers whose first language is not English yet who have enough command of that language to attempt reading the English literature. When reading technical literature in a foreign language, full explanations, such as those in the glossary, may not be necessary to convey the meaning of a few key words. For instance, a Spanish-speaking reader may need only the translation of a technical term from English to Spanish or even to a third language, rather than the extensive English explanation given in the glossary.

The dictionaries arranged by entries other than English serve two purposes depending on whether or not the user's main language is English. For English speakers reading in a foreign language, the dictionaries by entries other than English will serve the opposite purpose as that mentioned earlier. If the problem in understanding a term is not one of language, then the English reader can proceed further into the glossary and find a definition. On the other hand, for users fluent in languages other than English who wish to read the geostatistical literature in their native languages, a dictionary by entry other than English provides pointers to the English words in the glossary. Of course, such readers will have to have an elementary command of English for the glossary to be useful.

While the committee hopes that it has created a balanced and complete work, we realize that this glossary is open-ended and can never be truly completed. To continue to refine the list of terms, definitions, and translations would have certainly delayed making this work available. Therefore, despite the remaining possible shortcomings, we have proceeded to press. Comments on definitions as well as discussions on any part of this glossary are welcomed by the editor. Information concerning possible errors, omissions, and inadequacies will be particularly appreciated.

This dictionary and glossary would have not been possible without help from many individuals. The Chinese dictionary is the end product of a combined effort by Dr. Zhou, South-China Sea Institute of Oceanography; Prof. Jingru Hou, Beijing Institute of Science and Technology; Prof. Rendo Wang, China University at Wuhan; and Hua Zhu, Stanford University. Prof. Dr. Ing. H. Siemes, Institut für Mineralogie und Lagerstättenlehre, RWTH Aachen, and Dr. Helmut Schaeben, at the time with Geologisches Institut, Universität Bonn, helped in the preparation of the German portion of the dictionary. Dr. Barbara Namyslowska-Wilczyńska, Geotechnics Institute, Technical University of Wroclaw, Poland, produced the Russian section which benefited from suggestions by Dr. Arkady N. Bugaets, The USSR Ministry of Geology, Alma-Ata, USSR. Prof. Fernando Muge, Instituto Superior Técnico, Lisbon, wrote the Portuguese dictionary. Prof. J. Samper C., Escola Tècnica Superior d'Enginyers de Camins, Canals i Ports, Barcelona, reviewed the Spanish section. Prof. John C. Davis, Kansas Geological Survey, and anonymous reviewers made valuable suggestions that helped to improve earlier versions of the glossary. Jo Anne DeGraffenreid, Kansas Geological Survey, prepared the final camera-ready copy for the printer. Illustrations were drafted by Renate E. Hensiek and C. Brian Forgey, Kansas Geological Survey.

Lawrence, Kansas
August 1990

R.A.O.

CONTENTS

PART I

THE GLOSSARY

TERMINOLOGY

ABSOLUTE DISPERSION. See **logarithmic semivariogram**.

ACCUMULATION. The product of a **grading** operation by the length of the grading **support** L.

ACCUMULATION POINT. (a) An accumulation point of a set of points S is a point P such that there is at least a member of the set S distinct from P in any **neighborhood** of the given point. The accumulation point may or may not be a member of S. The number 0 is the accumulation point of the set $\{1,\frac{1}{2},\frac{1}{3},\ldots,\frac{1}{n},\ldots\}$. (b) An accumulation point of a sequence is a point P such that, for any integer n, each neighborhood of P contains at least one term of the sequence after the nth term. For example, the sequence $1,\frac{1}{2},1,\frac{1}{3},\ldots,1,\frac{1}{n},\ldots$ has two accumulation points, the numbers 0 and 1.

ACCURACY. The quality of an observation or a **statistic**, derived from a number of observations, of being free from **error**. Accuracy denotes the ultimate excellence of **data** and computed results. Accuracy should be distinguished from **precision**, which is only a factor of accuracy. Accuracy presumes precision, but the converse is not true. For instance, as a value of π, the number 3.1313 is more precise than accurate. There is not any standard **measure** for accuracy; the most common yardstick is the **mean square error**.

ADDITIVE SET FUNCTION. Let A and B be any two sets that do not share any elements. A **function** $f(\cdot)$ is said to be an additive set function if

$$f(A \cup B) = f(A) + f(B)$$

The **accumulation** of some **regionalized variables**, such as the amount of metal, produces a new regionalized variable that is additive; however not all regionalized variables are additive. Permeability is not an additive regionalized variable. The additivity property is a necessary condition for a regionalized variable to be **estimated** by **block kriging**.

ADDITIVITY THEOREM. (a) A synonym for **Krige's formula**. (b) A proposition that states that an **estimation** made by **universal kriging** is identical to the sum of two terms, the optimal **estimate** of the **drift**, and the estimate by **simple kriging** of the **optimal residuals**.

AFFINE CORRECTION. The simplest of various **support effect** correction procedures. The basic idea is to reduce the **variance** of the **distribution** by moving all values closer to the **mean**. The affine correction **transforms** a **quantile** $q_s(p)$ of the distribution for the smaller **support** into a quantile $q_l(p)$ associated with a larger support using the following linear formula

$$q_l(p) = m + \sqrt{f}\,[q_s(p) - m]$$

where m is the constant mean for both distributions and f is the **variance adjustment factor**. The affine correction does not account for the change in symmetry as it preserves the shape of the original distribution. Its main disadvantage is that it produces a minimum value that may not be realistic.

AFFINITY TRANSFORMATION. A linear **transformation** of an original **vector** $\boldsymbol{x}$ of n coordinates not involving any rotation. The transformation is usually performed in order to **model** a **geometric anisotropy** or to set 1 or up to $(n-1)$ coordinates to 0. If $\boldsymbol{x}'$ represents the transformed coordinate vector, then the affinity transformation is given by $x' = Ax$, where A is a **diagonal matrix**. The **order of the square matrix** is n.

ALEATORY VARIABLE. A rarely used synonym for **random variable**.

ANAMORPHOSIS. A strictly monotonic **transformation** of one **function** into another. **Geostatistical** anamorphoses are performed to transform one **distribution function** into another, such as the **affine correction**, the **indirect lognormal correction**, and the **Gaussian anamorphosis**.

ANISOTROPY. The quality or state of not possessing **isotropy**.

APPARENT DISCONTINUITY AT THE ORIGIN. A synonym for **nugget effect**.

A PRIORI VARIANCE. The **variance**, if it exists, of a **random function**. This is the limit of the **dispersion variance** of the random function within an infinitely large **domain**. There are random function **models** in which the dispersion variance increases with increasing domain size. In these models, the a priori variance is undefined.

ARITHMETIC AVERAGE. A synonym for **arithmetic mean**.

ARITHMETIC MEAN. The arithmetic mean of n numbers is

$$\bar{x} = \frac{1}{n} \sum_{i=1}^{n} x_i$$

The arithmetic mean is a sample **estimate** of the **expected value**. Provided not all observations in the **sample** are identical, the arithmetic mean of positive numbers is always larger than the **geometric mean** or the **harmonic mean**. See also **ergodicity**.

AUTHORIZED COEFFICIENTS OF ORDER v. A synonym for **permissible coefficients of order v**.

AUTHORIZED COMBINATIONS OF ORDER v. A synonym for **permissible combinations of order v**.

AUTOCORRELATION. The **correlation** of a **random process** with itself or between members of a series of observations ordered in time or space.

AUTOCORRELATION COEFFICIENT. The ratio of the **autocovariance** and the **variance** of a **random process**.

AUTOCORRELATION FUNCTION. The series resulting for all possible **lags** τ for the **autocorrelation coefficient**. A graph of the autocorrelation function versus lag is called an **autocorrelogram**.

AUTOCORRELOGRAM. The graphical display of an **autocorrelation function**.

AUTOCOVARIANCE. Let $\{Y(t_i):t_i \in T\}$ be a **random process** and let $Y(t_1)$ and $Y(t_2)$ be any two **random variables** in the series which are separated by a **lag** τ. The autocovariance is the **covariance** between $Y(t_1)$ and $Y(t_2)$. The term autocovariance is seldom applied to spatial processes; the more generic term covariance is used instead.

AUTOKRIGEABILITY. Given a **coregionalization**, one or more of the **random functions** possess the property of autokrigeability if the **autokriging** of any of those functions provides the same result as **cokriging**. An obvious coregionalization having autokrigeability is a collection of independent random functions defined over the same **field**. For the particular case of a coregionalization of two **random variables** whose random functions both possess autokrigeability, the two random functions are either **stochastically independent** or possess **intrinsic correlation**.

AUTOKRIGING. **Kriging** of only one **random function** of a **coregionalization**. It is used in opposition to **cokriging**, which uses the information for all **regionalized variables**.

AVERAGE. A familiar but elusive expression of centrality of a set of values. (a) Generally, an average is intended to represent or to summarize the relevant features of the set, and in this sense the term includes the **median** and the **mode**. (b) In ordinary usage, the average often refers to the **arithmetic mean**.

AVERAGE SEMIVARIOGRAM. A synonym for **mean semivariogram**.

AVERAGE VALUE. The average value of a **function** $f(x)$ is the integral $\frac{1}{S}\int_S f(x)dx$, where $f(x)$ is an **integrable function** in a portion S of its **domain**.

BASIS FUNCTION. (a) Any **function** used to **model** the **drift** in the **universal kriging** system of equations. In practice, the choice is limited to polynomials and more rarely to trigonometric functions, but in theory any **linearly independent** functions could be used. (b) In **intrinsic kriging**, those functions that are **filtered** by the **permissible coefficients of order** $\boldsymbol{v}$. A theoretical constraint forces the basis function $f(x_1,x_2,\ldots,x_i,\ldots,x_n)$ to be of the form

$$f(x_1,x_2,\ldots,x_i,\ldots,x_n)=p(x_1,x_2,\ldots,x_i,\ldots,x_n)e^{\sum_{i=1}^{n}a_i x_i}$$

where $p(x_1,x_2,\ldots,x_i,\ldots,x_n)$ is a polynomial in the geographical coordinates and a_i is a real or complex constant number. This model includes, as a special case, polynomials, exponentials, and trigonometric functions.

BAYES' THEOREM. Suppose $\mathcal{A}$ is an **event** and $\{\mathcal{B}_1,\mathcal{B}_2,\ldots,\mathcal{B}_i,\ldots,\mathcal{B}_n\}$ is a **partition** of a **sample space**. If $P(\mathcal{A})>0$, then the **conditional probability** of $\mathcal{B}_j$ given that $\mathcal{A}$ has occurred, $P(\mathcal{B}_j|\mathcal{A})$, is given by

$$P(\mathcal{B}_j|\mathcal{A})=\frac{P(\mathcal{B}_j)P(\mathcal{A}|\mathcal{B}_j)}{\sum_{i=1}^{n}P(\mathcal{B}_i)P(\mathcal{A}|\mathcal{B}_i)}$$

Suppose that four urns are equally likely to be **sampled**. Number 1 contains one white and two red balls; number 2 has three white and one red; number 3 has one white and four red; and number 4 has one white and five red. Then, the **probability** $P(\mathcal{B}_i)$ of an urn being sampled is $\frac{1}{4}$; and if $\mathcal{A}$ is the draw of a white ball, $P(\mathcal{A}|\mathcal{B}_i)$ equals $\frac{1}{3}, \frac{3}{4}, \frac{1}{5}$, and $\frac{1}{6}$, respectively, for each urn. According to Bayes' theorem, the probability $P(\mathcal{B}_3|\mathcal{A})$ that the white ball will come from urn number 3 is

$$P(\mathcal{B}_3)=\frac{\frac{1}{4}\cdot\frac{1}{5}}{\frac{1}{4}\cdot\frac{1}{3}+\frac{1}{4}\cdot\frac{3}{4}+\frac{1}{4}\cdot\frac{1}{5}+\frac{1}{4}\cdot\frac{1}{6}}=\frac{12}{87}$$

Bayes' theorem can also be formulated in terms of **probability density functions** of unknown parameters of a **distribution**. According to the Bayesian view, all quantities are of two kinds: those known to the person making the inference and those unknown to such person. The former situation is described by their known values, the **uncertainty** surrounding the latter being described by a **joint probability density function**. The statistical **model** admits a distribution of the **random variable** X for a parameter θ. When X has a probability density function or **like-**

lihood function $f(x|\theta)$, and $g(\theta)$ is the prior probability density function of θ, the posterior probability density function of θ, $g(x|\theta)$, is

$$g(\theta|x)=\frac{f(x|\theta)g(\theta)}{\int_\theta f(x|\theta)g(\theta)d\theta}$$

Both θ and x may be **vectors**.

BERNOULLI EXPERIMENT. A **trial** for which there are only two possible **outcomes**, such as obtaining a head or a tail when tossing a coin. Repetition of the **experiment** results in **stochastically independent** outcomes with constant **probability**. The **distribution** of successes in a sequence of trials is given by the **binomial distribution**.

BESSEL INEQUALITY. Let $\boldsymbol{u}$ be a **vector** of length $\|\boldsymbol{u}\|$ in an n-dimensional **euclidean space**, and let $\{\boldsymbol{u}_1,\boldsymbol{u}_2,\ldots,\boldsymbol{u}_i,\ldots,\boldsymbol{u}_m\}$ be a set of **orthonormal vectors**. Then, the Bessel inequality states that

$$\sum_{i=1}^{m}\langle \boldsymbol{u},\boldsymbol{u}_i\rangle^2\le\|u\|^2$$

where $\langle \boldsymbol{u},\boldsymbol{u}_i\rangle$ denotes a **scalar product**. For the particular case $m=n$, the equality holds and is called the *Pythagorean theorem* or *Parseval's equation*. Equality also holds for infinite orthonormal vectors in a **Hilbert space**. If the scalar product is defined by integration, then

$$\langle \boldsymbol{u},\boldsymbol{u}_i\rangle=\int_{\Re}f(x)g_i(x)dx$$

and the Bessel inequality becomes

$$\sum_{i=1}^{m}[\int_{\Re}f(x)g_i(x)dx]^2<\int_{\Re}[f(x)]^2dx$$

BETA DISTRIBUTION. Let X be a **random variable** whose **range** is the interval (0,1), and let α and β be two positive numbers. The random variable X is beta-distributed if its **probability density function** is given by

$$f(x)=\frac{1}{B(\alpha,\beta)}x^{\alpha-1}(1-x)^{\beta-1},\quad 0<x<1$$

where $B(\cdot)$ is the **beta function**. The **mean** of the beta distribution is $\alpha/(\alpha+\beta)$ and its **variance** is $\alpha\beta/[(\alpha+\beta)^2(\alpha+\beta+1)]$.

BETA FUNCTION. Let m and n be positive numbers. The beta function $B(m,n)$ is the integral

$$B(m,n)=\int_0^1 t^{m-1}(1-t)^{n-1}dt$$

It can be shown that

$$B(m,n)=\frac{\Gamma(m)\Gamma(n)}{\Gamma(m+n)}$$

where $\Gamma(\cdot)$ is the **gamma function**.

BIAS. In its broadest sense, a bias is any effect that deprives a statistical result of representativeness by systematically distorting it. In contrast, a random **error** may introduce distortion occasionally but balances out on the **average**. For a given **estimator** Φ of a parameter ϕ, bias is the **expected value** $\mathrm{E}[\Phi+\phi]$ when the expectation is not 0. Bias may arise from the specific form of the estimator, poor selection of **samples**, or systematic errors of measurement.

BIGAUSSIAN MODEL. A stochastic **model** of a **stationary** type of **random function** such that any pair of **random variables** $Z(\boldsymbol{x})$ and $Z(\boldsymbol{x}+\boldsymbol{h})$ follow a **bivariate normal distribution** for any value of $\boldsymbol{h}$. The **probability density function** of this bivariate normal distribution can be factored into an **isofactorial bivariate density function** in which the parameter r is given by

$$r=\frac{\mathrm{Cov}[Z(\boldsymbol{x}),Z(\boldsymbol{x}+\boldsymbol{h})]}{\mathrm{Var}[Z(\boldsymbol{x})]}$$

BIJECTION. One-to-one correspondence between two sets A and B. Each element of A is paired with exactly one element of B.

BIMODAL DISTRIBUTION. A **distribution** having two **modes**.

BINOMIAL DISTRIBUTION. Let X be a **discrete random variable** that can assume any nonnegative integer value that is no larger than n, and let p be a number larger than 0 but less than 1. The random variable X is binomially distributed if its **probability function** is given by

$$P(X=x)=\frac{n!}{x!(n-x)!}p^x(1-p)^{n-x}$$

The **mean** of the binomial distribution is np and its **variance** $np(1-p)$. In a **Bernoulli experiment**, the binomial distribution **models** the number of occurrences of an **outcome** whose **probability** is p in a series of n independent **trials**. As n tends to infinity, the binomial distribution approximates the **normal distribution**.

BIVARIATE. Pertaining to two variables that are considered simultaneously. Bivariate **probability density functions** are especially important in **geostatistics**. Although in theory many studies depend on **multivariate distributions**, in practice analysis of the simpler properties of **bivariate distributions** is often sufficient.

BIVARIATE DISTRIBUTION. The **joint probability distribution**, $F(z_1,z_2)$, of two **random variables** Z_1 and Z_2

$$F(z_1,z_2)=P(Z_1\leq z_1,Z_2\leq z_2)$$

For a **random function** $Z(\cdot)$ that is **stationary**, the joint distribution can be interpreted as the **expected value** of the product of two **indicators** $I(\boldsymbol{x},z_i)$ and $I(\boldsymbol{x}+\boldsymbol{h},z_j)$

$$F(\boldsymbol{h};z_i,z_j)=\mathrm{E}[I(\boldsymbol{x},z_i)\,I(\boldsymbol{x}+\boldsymbol{h},z_j)]$$

Thus, a bivariate distribution function can be inferred from **indicator cross-covariances**.

BIVARIATE NORMAL DISTRIBUTION. The **joint probability density function** $f(x,y)$ for two **random variables** X and Y given by

$$g(x,y)=\frac{1}{2\pi\sigma_X\,\sigma_Y\sqrt{1-r^2}}\,e^{-\frac{0.5w}{1-r^2}}$$

where μ_X and μ_Y are the **means** of X and Y; σ_X and σ_Y are their **standard deviations**; r is the **correlation coefficient** between X and Y; and w is given by

$$w=\left(\frac{x-\mu_X}{\sigma_X}\right)^2-2r\frac{(x-\mu_X)(y-\mu_Y)}{\sigma_X\sigma_Y}+\left(\frac{y-\mu_Y}{\sigma_Y}\right)^2$$

The distribution has several appealing properties over other bivariate distributions. In particular, the conditional **variances** are **homoscedastic**, the **regression** is linear, any **linear combination of variables** in pairs is also a **normal distribution**, and the **conditional probability density function** as well as the **marginal probability density functions** are normal. However, two normally distributed variables do not necessarily have a joint bivariate normal distribution. A normal distribution in more than two random variables is called a **multinormal distribution**.

BLOCK KRIGING. **Estimation** by **kriging** of **average values** defined over a **support** larger than the support for the observations. The block kriging system of equations is the same as for kriging when the support of the **estimate** is the same as the support of the observations, except that the right-hand side of the **structural function** for the **sample** support is replaced by its average value between the block and the observation support. Note that the average of an attribute over a given volume for a non**additive set function** may be an irrelevant value. For example, averages of permeabilities measured on small pieces of core are not estimates of permeabilities actually measured on the whole core.

BOOLEAN RANDOM FUNCTION. Let $\Re^n$ be an n-dimensional **euclidean space** in which a **Poisson process** takes place. Let us also define random **supports** centered at these points. The nature of their volumes does not depend on the location of the points and overlapping is allowed. A Boolean random function is the union of the external, nonoverlapping portions of these volumes in $\Re^{n+1}$. These kinds of functions have been used in the **simulation** of **regionalized variables**.

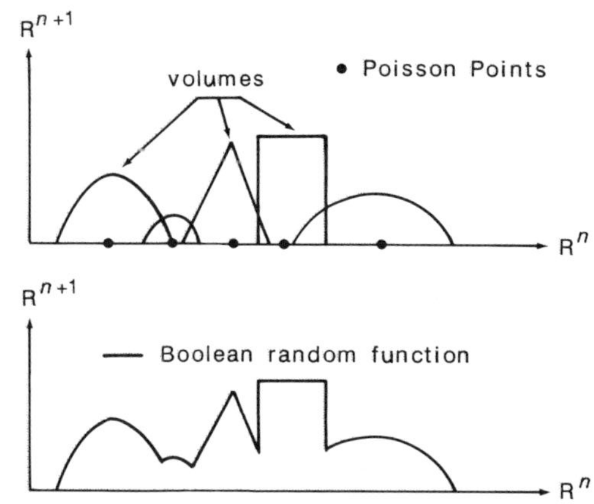

Copyright source: Chautru, 1989

BOREL FIELD. Any set that can be obtained by a countable number of operations, starting from **open** sets. The operations may consist of unions, intersections, or complements. The collection of all Borel sets in a **euclidean space** of dimension n is the smallest **σ-algebra** that contains all open sets.

BROWNIAN MOTION PROCESS. An application of the **Wiener process** to describe thermal noise and the movements of microscopic particles immersed in a fluid.

BUNIAKOVSKI INEQUALITY. An alternative term for **Schwarz inequality**.

CARTIER'S RELATION. Cartier's relation states that, for two **random variables** X and Y,

$$\mathrm{E}[X|Y] = Y$$

The implication of the relation is that the **distribution function** of Y is less **selective** than the distribution of X and the **variance** of Y is smaller than the variance of X. An important example relates to the **change of support model**. Let Z_v be the **average value** of the **random function** $Z(\boldsymbol{x})$ for the **support** v, let $\bar{x}$ be a location inside v selected at random in a way that any location has the same **probability** of being selected, and let $Z_{\bar{x}}$ be the value of $Z(\boldsymbol{x})$ at location $\bar{x}$. Then

$Z_{\bar{x}}$ and Z_v conform to Cartier's relation

$$E[Z_{\bar{x}} | Z_v] = Z_v$$

CAUCHY–SCHWARZ INEQUALITY. See **Schwarz inequality.**

CAUCHY SEQUENCE. A sequence with the property that the difference between any two terms is arbitrarily small provided that both are sufficiently separated in the sequence. More precisely, a sequence $\{a_1, a_2, \ldots, a_i, \ldots, a_j, \ldots\}$ such that for any $\varepsilon > 0$, there is an integer N with the property that if i and j are both greater than N, then $\|a_j - a_i\| < \varepsilon$. If the terms belong to a **euclidean space**, then this is equivalent to the sequence being convergent.

CENTERED COVARIANCE. A synonym for **covariance.**

CENTILE. A rarely used abbreviated form of **percentile.**

CENTRAL LIMIT THEOREM. Any of the theorems that provide the mathematical formal conditions that sustain the **experimental** evidence that the **distribution** of **sample** means usually approximate a **normal distribution**, regardless of the nature of the **distribution function** of the parent **population**. For example, if all the **random variables** in a sequence $\{X_1, X_2, \ldots, X_i, \ldots, X_n\}$ are stochastically independent and have the same distribution function with finite **mean** μ and **variances** σ^2, then the distribution of

$$\frac{\sum_{i=1}^{n} X_i - \mu}{\sigma / \sqrt{n}}$$

approaches a **standard normal distribution** as n tends to infinity. The central limit theorem justifies the wide use of the normal **model** to represent the distribution of a sum of additive effects possessing **stochastic independence**, such as instrumental **errors**, regardless of the **probability function** of the individual effects. The independence may be violated in many natural phenomena, thus accounting for the failure of many spatial processes to follow normal distributions.

CHANGE OF SUPPORT MODEL. A **model** for the **bivariate distributions** between any two linear **average values** defined on **supports** that may be different and centered at different locations. If **Cartier's relation** is used to establish such a model, one of the average values must be defined on the smaller support and be located randomly within the support of the second average value.

CHARACTERISTIC EQUATION OF A MATRIX. Let $\boldsymbol{A}$ be a **square matrix** and let $\boldsymbol{B}$ be the **matrix** that results from subtracting an unknown constant λ from each element of the **main diagonal of matrix** $\boldsymbol{A}$. The characteristic equation of matrix $\boldsymbol{A}$ is the equation

$$|\boldsymbol{B}| = 0$$

where $|\boldsymbol{B}|$ is the **determinant of matrix** $\boldsymbol{B}$. The characteristic equation is a **polynomial of degree** $\boldsymbol{n}$. For instance, if

$$\boldsymbol{A} = \begin{bmatrix} 4 & 2 & 7 \\ 0 & -1 & 5 \\ 3 & 6 & 1 \end{bmatrix}$$

then its characteristic equation is

$$\begin{vmatrix} (4-\lambda) & 2 & 7 \\ 0 & -(\lambda+1) & 5 \\ 3 & 6 & (1-\lambda) \end{vmatrix} = -\lambda^3 + 4\lambda^2 + 52\lambda - 73 = 0$$

CHOLESKI'S METHOD. An alternative term for **lower-upper (LU) decomposition.**

CHOQUET'S CAPACITY. See **random set.**

CLOSED. The property of a set of points denoting that all its **accumulation points** belong to the set.

COCONDITIONAL SIMULATION. The **simulation** of a spatial **coregionalization.** Although a form of **multivariate** simulation can be performed by simulating one **random function** at a time using **kriging** for the conditioning of each variable, this may result in an incorrect portrayal of the **cross-correlation** between the simulated random functions. This is particularly true if one or more of the variables are sparsely **sampled.** This shortcoming is avoided by the use of **cokriging** for the conditioning.

CODIMENSION. The difference between the dimension of a hyperplane and the dimension of the **euclidean space** in which the hyperplane is defined. For instance, the codimension of a two-dimensional plane in a three-dimensional space is 1.

COEFFICIENT OF EXCESS. A **measure** of **kurtosis** defined by

$$B_2 - 3$$

where B_2 is the **coefficient of kurtosis.**

COEFFICIENT OF KURTOSIS. A **measure** of peakedness in a **distribution,** given by the ratio

$$B_2 = \frac{u_4}{u_2^2}$$

where u_2 and u_4 are the second and the fourth **moments** about the **mean.** A **normal distribution** has $B_2 = 3$, and is said to be *mesokurtic.* A **distribution** with $B_2 < 3$ is *platykurtic,* with a lower concentration of values near the mean then a normal distribution. If $B_2 > 3$, the distribution is *leptokurtic* and has an excess of values near the mean.

COEFFICIENT OF SKEWNESS. A measure of asymmetry in a **distribution,** given by the ratio

$$B_1 = \frac{u_3}{u_2^{3/2}}$$

where u_2 and u_3 are the second and the third **moments** about the **mean.** A **distribution** with a tail extending to the left has a **median** larger than the mean and a negative skewness coefficient; symmetric distributions, like the **normal distribution** in which the mean and the median coincide, have a coefficient of 0, but the converse is not true; and distributions, like the **lognormal distribution,** with a tail extending to the right, have a median smaller than the mean and a positive coefficient. The coefficient of skewness suffers even more than the mean and the variance from a sensitivity to **outliers.**

COEFFICIENT OF VARIATION. For a **distribution** of positive **outcomes,** the quotient of the **standard deviation** over the **mean.** This parameter is used as a **measure** of relative dispersion.

COFACTOR OF AN ELEMENT IN A DETERMINANT. Let

$$\begin{vmatrix} a_{11} & a_{12} & \cdots & a_{1(j-1)} & a_{1j} & a_{1(j+1)} & \cdots & a_{1n} \\ a_{21} & a_{22} & \cdots & a_{2(j-1)} & a_{2j} & a_{2(j+1)} & \cdots & a_{2n} \\ \vdots & \vdots & \cdots & \vdots & \vdots & \vdots & \cdots & \vdots \\ a_{(i-1)1} & a_{(i-1)2} & \cdots & a_{(i-1)(j-1)} & a_{(i-1)j} & a_{(i-1)(j+1)} & \cdots & a_{(i-1)n} \\ a_{i1} & a_{i2} & \cdots & a_{i(j-1)} & a_{ij} & a_{i(j+1)} & \cdots & a_{in} \\ a_{(i+1)1} & a_{(i+1)2} & \cdots & a_{(i+1)(j-1)} & a_{(i+1)j} & a_{(i+1)(j+1)} & \cdots & a_{(i+1)n} \\ \vdots & \vdots & \cdots & \vdots & \vdots & \vdots & \cdots & \vdots \\ a_{n1} & a_{n2} & \cdots & a_{n(j-1)} & a_{nj} & a_{n(j+1)} & \cdots & a_{nn} \end{vmatrix}$$

be a **determinant** for which the **order of the determinant** is larger than 2. The cofactor $|C_{ij}|$ of an element a_{ij} is the signed determinant of the matrix of next lower order, obtained by striking out row i and column j in which the element a_{ij} lies

$$|C_{ij}| = (-1)^{i+j} \begin{vmatrix} a_{11} & a_{12} & \cdots & a_{1(j-1)} & a_{1(j+1)} & \cdots & a_{1n} \\ a_{21} & a_{22} & \cdots & a_{2(j-1)} & a_{2(j+1)} & \cdots & a_{2n} \\ \vdots & \vdots & \cdots & \vdots & \vdots & \vdots & \cdots \\ a_{(i-1)1} & a_{(i-1)2} & \cdots & a_{(i-1)(j-1)} & a_{(i-1)(j+1)} & \cdots & a_{(i-1)n} \\ a_{(i+1)1} & a_{(i+1)2} & \cdots & a_{(i+1)(j-1)} & a_{(i+1)(j+1)} & \cdots & a_{(i+1)n} \\ \vdots & \vdots & \cdots & \vdots & \vdots & \vdots & \cdots \\ a_{n1} & a_{n2} & \cdots & a_{n(j-1)} & a_{n(j+1)} & \cdots & a_{nn} \end{vmatrix}$$

COKRIGING. Estimation of a **regionalized variable** using not only observations on the variable, but also **data** on one or more additional related regionalized variables defined over the same **field.** Cokriging can be regarded as a generalization of **kriging** in the sense that at every sampling location, there is a **vector** $[Z_1(\boldsymbol{x}_i), Z_2(\boldsymbol{x}_i), \ldots, Z_j(\boldsymbol{x}_i), \ldots, Z_m(\boldsymbol{x}_i)]$, instead of a single value $\mathbf{Z}(\boldsymbol{x}_i)$. Cokriging can be used to reduce the **estimation variance** when the variable of interest has been undersampled, or for the simultaneous estimation of several regionalized variables **sampled** over the same field.

COLUMN OF A DETERMINANT. A collection of elements along a vertical line and in the same sequence as they appear in a **determinant.**

COLUMN OF A MATRIX. A collection of elements along a vertical line and in the same sequence as they appear in a **matrix.**

COMPACT. A subset A of a **metric space** is said to be compact if every sequence in A contains a convergent sequence whose limit belong to A.

COMPLETE SET OF ORTHOGONAL FUNCTIONS. Let $f(x)$ be a square **integrable function with respect to the probability density function** $g(x)$. A square integrable set of functions with respect to $g(x)$, $\{q_0(x), q_1(x), q_2(x), \ldots\}$, is complete if and only if the **Parseval's equation**

$$\int_{\Re} f^2(x) g(x) dx = \sum_{k=0}^{\infty} f_k^2$$

is valid. The f_k are expansion coefficients

$$f_k = \mathrm{E}[f(x) q_k(x)]$$

with respect to the density $g(x)$.

COMPLETE SET OF ORTHOGONAL POLYNOMIALS. A special case of a **complete set of orthogonal functions** when the orthogonal functions are polynomials.

COMPLETE SPACE. A **metric space** in which every **Cauchy sequence** converges to a point of the space.

CONDITIONAL BIAS. Let Z be a **random variable** and Z^* an **estimator**. The conditional bias of Z^* is the **expected value**

$$\mathrm{E}[Z - Z^* | Z^* = z] = \mathrm{E}[Z | Z^* = z] - z$$

where z is any given value.

CONDITIONAL COSIMULATION. An alternative term for **coconditional simulation**.

CONDITIONAL DISTRIBUTION. The **distribution function** of one or more **random variables** conditional on **realizations** or on constraints on one or more related random variables.

CONDITIONAL EXPECTATION. The first order **moment** of a **conditional distribution**.

CONDITIONALLY POSITIVE DEFINITE OF ORDER ν. A continuous, symmetric **function** $f(x_i - x_j)$ is said to be conditionally positive definite if

$$\sum_{i=1}^{m} \sum_{j=1}^{m} a_i a_j f(x_i - x_j) \geq 0$$

for all **permissible coefficients of order ν**, $\{a_1, a_2, \ldots, a_i, \ldots, a_m\}$. In **geostatistics**, the function $f(x_i - x_j)$ is either a **geometric covariogram**, a **generalized covariance**, or the negative of the **semivariogram**.

CONDITIONAL NONBIAS. A synonym for **conditional unbiasedness**.

CONDITIONAL PROBABILITY. For two **events** $\mathcal{A}$ and $\mathcal{B}$, the latter with nonzero **probability**, the conditional probability of $\mathcal{A}$ given $\mathcal{B}$ is the probability of $\mathcal{A}$, assuming $\mathcal{B}$ holds. Conditional probability is denoted by the symbol $P(\mathcal{A} | \mathcal{B})$, and is defined by

$$P(\mathcal{A} | \mathcal{B}) = \frac{P(\mathcal{A} \cap \mathcal{B})}{P(\mathcal{B})}$$

Bayes' theorem makes use of conditional probabilities.

CONDITIONAL PROBABILITY DENSITY FUNCTION. The **probability density function** resulting when values for one or more **random variables** are given. A general expression for the conditional probability density function of k variables $f_1(x_1, x_2, \ldots, x_k | x_{k+1}, x_{k+2}, \ldots, x_n)$, given the remaining $(n-k)$ variables is

$$f_1(x_1,x_2,\ldots,x_k|x_{k+1},x_{k+2},\ldots,x_n)=\frac{f_1(x_1,x_2,\ldots,x_n)}{f_2(x_{k+1},x_{k+2},\ldots,x_n)}$$

where $f(x_1,x_2,\ldots,x_n)$ is the **joint probability density function** for all n variables, and $f_2(x_{k+1},x_{k+2},\ldots,x_n)$ is the **marginal probability density function** for the $(n-k)$ variables $x_{k+1},x_{k+2},\ldots,x_n$ whose values are known.

CONDITIONAL SIMULATION. A **simulation** of a single **random function** honoring the **sample** values. Conditional simulation can be performed directly using a **lower-upper decomposition** of **covariance matrices** or the **sequential indicator simulation** algorithm, or indirectly by the **turning bands method**. The latter process combines three random functions and uses two sets of known values to force the simulation to honor the **data**. Let $z^*(\boldsymbol{x}_0)$ be a **kriging** estimate at location $\boldsymbol{x}_0$ of a **regionalized variable** defined over the n-dimensional **euclidean space** $\Re^n$, **modeled** by the random function $Z(\boldsymbol{x})$. This **estimate** is based on a set of m observations $z(\boldsymbol{x}_i)$. Let $z_1(\boldsymbol{x}_0)$ be a single **realization** of an unconditional simulation of $Z(\boldsymbol{x})$ produced by the turning bands method using the same original data. Let $z_2(\boldsymbol{x}_0)$ be a kriging estimate of $z(\boldsymbol{x}_0)$ calculated from a realization of an unconditional simulation of $Z(\boldsymbol{x})$ produced by the turning bands method at the same locations $\boldsymbol{x}_i$ as the original data set. A realization $z_3(\boldsymbol{x}_0)$ of the conditional simulation of $Z(\boldsymbol{x})$ at $\boldsymbol{x}_0$ is given by

$$z_3(\boldsymbol{x}_0)=z^*(\boldsymbol{x}_0)+z_1(\boldsymbol{x}_0)-z_2(\boldsymbol{x}_0)$$

Coconditional simulation denotes the joint conditional simulation of more than one random function.

CONDITIONAL UNBIASEDNESS. The property of a **random function** estimator $Z^*(\boldsymbol{x}_0)$ to produce **estimates** that are free of **bias**, regardless of the magnitude of the estimated value. Symbolically

$$\mathrm{E}[(Z^*(\boldsymbol{x}_0)-Z(\boldsymbol{x}_0))|Z^*(\boldsymbol{x}_0)=z]=0,\ \text{for all}\ z\in D$$

where $Z(\boldsymbol{x}_0)$ is a **random variable** at location $\boldsymbol{x}_0$, and z is any **realization** of the **estimator** $Z^*(\boldsymbol{x}_0)$ within its **range** D. This property is more stringent than the **global** constraint $E[Z^*(\boldsymbol{x}_0)-Z(\boldsymbol{x}_0)]=0$. An estimator possessing conditional unbiasedness is also globally **unbiased**, but the converse is not true. A **regression** curve of joint realizations of $Z^*(\boldsymbol{x}_0)$ and $Z(\boldsymbol{x}_0)$ must be a line coinciding with the main bisector.

CONFIDENCE INTERVAL. The interval between two **statistics** t_l and t_u, such that it is feasible to compute

$$P(t_l\le\theta\le t_u)=\alpha$$

where $P(\cdot)$ denotes the **probability** of the **event** in parentheses, θ is the parameter being **estimated**, and α is some fixed value. Customary probability levels for α are 95 and 99 percent.

CONFIDENCE LIMIT. Either of the two values t_l and t_u that define the lower and upper bounds of a **confidence interval.**

CONFIGURATION. A general term to denote the arrangement in any geometrical figure, or in any combination of geometrical elements such as points, lines, curves, and surfaces.

CONJUGATE COMPLEX NUMBER. The conjugate of a complex number $x=a+ib$, is denoted by $\bar{x}$ and is defined as another complex number having the same real part and an imaginary part with the reverse sign, namely $\bar{x}=a-ib$.

CONSERVATION OF LAW HYPOTHESIS. The presumption that a **stationary** type of **distribution function** is unvariant under **moving linear averaging**. **Random functions** that follow a **normal distribution** satisfy the conservation hypothesis.

CONSISTENCY. In a system, a lack of contradiction between parts or features to one another or as a whole. Consistency is a basic requirement of any reliable mathematical **model**.

CONSTRAINT EFFECT. In mining, a given block is not selected for extraction solely according to its **average** grade. Other considerations, such as technical factors, may require the mining of blocks that are below the cutoff grade or, more unlikely, the rejection of some rich blocks. As a result, the **mean** grade of produced ore is always lowered. This reduction in the average grade is called the constraint effect.

CONSTRAINT INTERVAL. A type of **soft data** in which the exact value of a **regionalized variable** $z(\mathbf{x})$ is not known but lies within an interval $]a,b]$. Such **data** may be coded and used in **indicator kriging** as an incomplete **indicator data column**. **Hard data** are but a particular case of constraint intervals for intervals of 0 length.

CONTINUOUS RANDOM VARIABLE. See **distribution function**.

CONTOUR. An alternative term for **isopleth**.

CONVENTIONAL PARAMETER. In a **model**, a parameter that does not satisfy the **objectivity criterion**. It is useful in intermediate steps that may be required in formulating a model, but should not be included in the final model. The **expected value** of a **stationary** type of **random function** is a conventional parameter, but the **local** value of the **mean** of a **regionalized variable** has an **objective meaning** and is not a conventional parameter.

CONVEX FUNCTION. Let λ be any real number between 0 and 1. The **function** $f(x)$ is convex if, for any pair of points x_1 and x_2,

$$f(\lambda x_1+(1-\lambda)x_2)\leq\lambda f(x_1)+(1-\lambda)f(x_2)$$

Geometrically, this is equivalent to stating that the line segment from $[x_1,f(x_1)]$ to $[x_2,f(x_2)]$ is above the curve.

CONVOLUTION. Let $f(x)$ and $g(x)$ be two **integrable functions**. Convolution is the mathematical operation denoted by $f(x)*g(x)$ and defined as

$$f(x)*g(x)=\int_0^x f(y)g(x-y)dy=\int_0^x g(y)f(x-y)dy$$

If $f(x)$ and $g(x)$ are the **probability density functions** of two **stochastically independent** continuous random variables X and Y, their convolution provides the probability density function of the sum of the two **continuous random variables**. Convolution is related to the operations of **filtering** and **regularization**.

COREGIONALIZATION. A set of two or more **regionalized variables** defined over the same **field**, such as the concentration of NO_3^- and Cl^- ions over an aquifer.

CORRELATION. The characteristic of having linear interdependence between **random variables** or between sets of numbers. For the particular case of two random variables, the degree of linear relationship is measured by the **correlation coefficient**.

CORRELATION COEFFICIENT. The ratio

$$\frac{\text{Cov}(X,Y)}{\sigma_X \sigma_Y}$$

where Cov(X,Y) is the **covariance** between **random variables** X and Y having **standard deviations** σ_X and σ_Y. The correlation coefficient varies between –1 and 1 and is a **measure** of the linear relationship between X and Y. A coefficient close to –1 indicates a tendency for the variables to vary inversely, so one variable increases as the other variable decreases. If both variables change in the same manner, the correlation coefficient is close to 1. A coefficient equal to 0 indicates lack of a linear relationship, but says nothing about other forms of interdependence.

CORRELOGRAM. A graphical display of the **correlation coefficient** between two **random variables** as a **function** of the **lag**. The term **autocorrelation function** is favored when a series is correlated with itself.

COSIMULATION. Simultaneous **simulation** of more than one **random function**.

COVARIANCE. The **expected value** $E[(X-\mu_x)(Y-\mu_y)]$ for pairs of **random variables** X and Y having a **joint probability density function**. The terms μ_x and μ_y represent the **mean** for X and Y. The covariance Cov(·) is a **measure** of linear relationship between pairs of random variables. When X and Y refer to the same **second order random field** $Z(\cdot)$ at two different locations $\boldsymbol{x}$ and $\boldsymbol{x}+\boldsymbol{h}$, the covariance becomes a **function** solely of the **lag** $\boldsymbol{h}$.

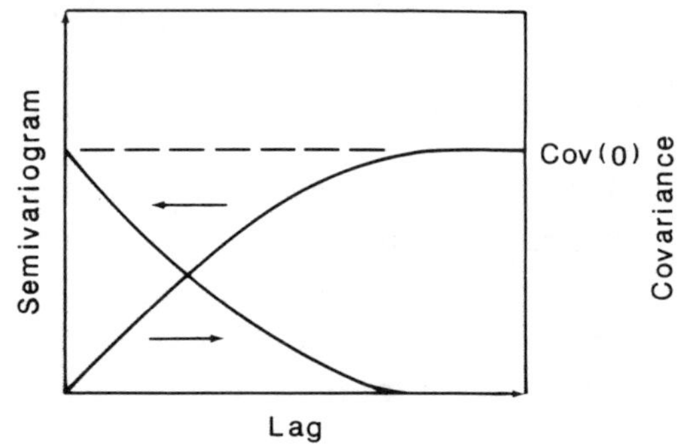

$$\text{Cov}(\boldsymbol{h}) = E[(Z(\boldsymbol{x})-m)(Z(\boldsymbol{x}+\boldsymbol{h})-m)] = E[Z(\boldsymbol{x})Z(\boldsymbol{x}+\boldsymbol{h})]-m^2$$

where m denotes a constant **drift**. The covariance is related to the **semivariogram**, $\gamma(\boldsymbol{h})$, through

$$\text{Cov}(\boldsymbol{h}) = \text{Cov}(0) - \gamma(\boldsymbol{h})$$

where Cov(0) is the **a priori variance**. In practice, the **population** covariance is approximated by the **experimental covariance**.

COVARIANCE MATRIX. The square array of n rows and n columns of elements $\text{Cov}(h_{ij})$, where Cov(·) denotes a **stationary** type of **covariance**, and h_{ij} is the distance between locations i and j, which are two positive integers not larger than n, the number of locations. This **matrix** is **positive definite**.

CROSS-CORRELATION or **CROSSCORRELATION**. **Correlation** under the particular circumstances that the two **random variables** being compared relate to different attributes.

CROSS-CORRELATION COEFFICIENT. A **correlation coefficient** when the two **random variables** involved in the comparison pertain to different attributes.

CROSSCORRELOGRAM. A **correlogram** for a pair of **random functions** that **model** two different attributes.

CROSS-COVARIANCE. A **covariance** in which the two **random variables** pertain to different attributes.

CROSS-PLOT or **CROSSPLOT**. A synonym for **scattergram**.

CROSS-SEMIVARIOGRAM. Let $Z_j(\boldsymbol{x}_i)$ and $Z_k(\boldsymbol{x}_i)$ be two **random variables** having a **joint probability distribution** that satisfies the **intrinsic hypothesis**. The cross-semivariogram $\gamma_{jk}(\boldsymbol{h})$ is a **measure** of their joint spatial variability defined as

$$\gamma_{jk}(\boldsymbol{h}) = 0.5\mathrm{E}[\{Z_j(\boldsymbol{x}_i+\boldsymbol{h}) - Z_j(\boldsymbol{x}_i)\}\{Z_k(\boldsymbol{x}_i+\boldsymbol{h}) - Z_k(\boldsymbol{x}_i)\}]$$

Unlike a **semivariogram**, a cross-semivariogram assumes negative values when increases in $Z_j(\boldsymbol{x}_i)$ correspond to reductions in $Z_k(\boldsymbol{x}_i)$. The cross-semivariogram is related to the **cross-covariance** $\mathrm{Cov}_{jk}(\boldsymbol{h})$ through

$$\gamma_{jk}(\boldsymbol{h}) = \mathrm{Cov}_{jk}(0) - 0.5[\mathrm{Cov}_{jk}(\boldsymbol{h}) + \mathrm{Cov}_{kj}(\boldsymbol{h})]$$

when the joint probability distribution is second order **stationary**.

CROSS-VALIDATION. Any of the **validation** methods in which observations are dropped one at a time from a **sample** of size n, and n **estimates** are computed using at most the remaining $(n-1)$ measurements. In principle the **estimation** should be performed reevaluating any parameter, such as the **semivariogram**, in whose estimation the discarded point might have been used. In practice, such parameters are supposed to be **robust** enough to be insensitive to the discarding of a single observation and the recalculation of parameters is never done. The comparison between the values calculated on the full sample and the censored subsets is the basis for evaluating the overall quality of the estimation procedure, namely the combined selection of the estimation method and its parameters. Search for better parameters can be done by trial and error or using **maximum likelihood** approaches.

CROSS-VARIOGRAM. A **function** equal to double the **cross-semivariogram**.

CUBIC SEMIVARIOGRAM. The **transitive** type of **semivariogram model**

$$\gamma(h) = \begin{cases} C\left(7\dfrac{h}{a} - \dfrac{35}{4}\left(\dfrac{h}{a}\right)^3 + \dfrac{7}{2}\left(\dfrac{h}{a}\right)^5 - \dfrac{3}{4}\left(\dfrac{h}{a}\right)^7\right) & ,0 \le h < a \\ C & ,a \le h \end{cases}$$

where C is the **sill** and a is the **range**.

CUMULATIVE DISTRIBUTION FUNCTION. A synonym for **distribution function**.

CUMULATIVE FREQUENCY. An alternative term for the **distribution function** of a **random variable** X.

CURVE FITTING. The process of finding an analytical expression that approximates a **data** set. The procedure involves two steps, the selection of the analytical expression and the evaluation of its constants. The most common curve fitting techniques are the **least squares method** using polynomials of low degree, and **interpolation** by **spline functions**. In some circumstances, as in **semivariogram fitting**, the curve is fitted iteratively by trial and error.

DATA. A synonym of meaning (a) for **sample**.

DATUM. (a) A surface or level to which measurements are referred, such as the sea level or the top of a stratigraphic unit. (b) Same as definition (b) for **sample**.

DECILE. See **quantile.**

DECLUSTERING. Any of the techniques intended to correct sampling bias caused by an excessive number of measurements within a limited portion of a **field**. If the sampled process is spatially correlated and there are areas that on **average** have a much higher observation density than others, any **statistic** computed from that original **sample** may have a significant **bias**. Declustering must be performed prior to any **structural analysis**.

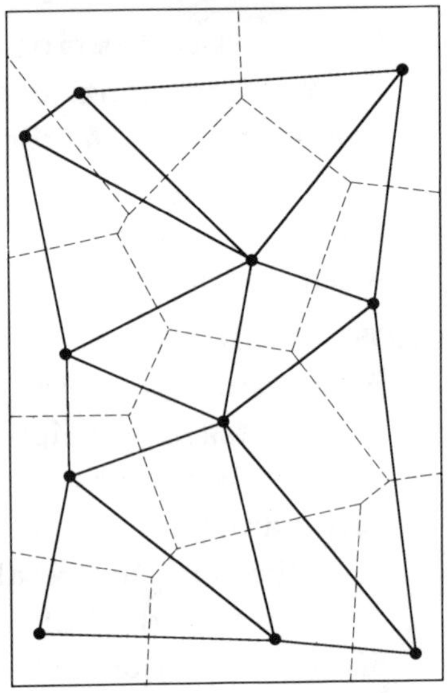

DECONVOLUTION. The process of undoing a previous **convolution**.

DELAUNAY SUBDIVISION. The graph obtained by joining pairs of points whose polyhedra in a **Voronoi division** have an $(n-1)$ dimensional face in common. The two-dimensional **Delaunay triangulation** is the primary case of practical interest.

DELAUNAY TRIANGULATION. A two-dimensional **Delaunay subdivision**. A Delaunay triangulation is uniquely defined for a given set of points. The triangulation is used as a basis for automatic **contouring. Interpolation** is done using only the three points at the vertices of the triangle enclosing the **estimation location**. The method works better for **configurations** that are almost regular and cannot be used for **extrapolations**.

DEMIVARIOGRAM. A rarely used synonym for **semivariogram**.

DETERMINANT. A square array of elements symbolizing a certain sum of products of these elements. The notation for a determinant is the same as for a **matrix** except that the elements are displayed inside vertical bars

$$|\mathbf{A}| = \begin{vmatrix} a_{11} & a_{12} & \cdots & a_{1n} \\ a_{21} & a_{22} & \cdots & a_{23} \\ \cdot & \cdot & \cdot & \cdot \\ a_{n1} & a_{n2} & \cdots & a_{nn} \end{vmatrix}$$

Unlike matrices, a determinant has a single scalar value. For a determinant of order 2, the determinant is simply

$$\begin{vmatrix} a_{11} & a_{12} \\ a_{21} & a_{22} \end{vmatrix} = a_{11}a_{22} - a_{12}a_{21}$$

For orders higher than 2, the expression represented by a determinant is the sum of the product of elements in a row or in a column by their corresponding **cofactors of the elements in the determinant** $|\mathbf{C}_{ij}|$

$$|\mathbf{A}| = \sum_{i=1}^{n} a_{ik} |\mathbf{C}_{ik}| = \sum_{j=1}^{n} a_{lj} |\mathbf{C}_{lj}|$$

The following example evaluates a determinant using as coefficients a_{ik} the elements of the second column

$$\begin{vmatrix} 4 & 5 & 6 \\ 1 & 9 & 2 \\ 3 & 0 & 7 \end{vmatrix} = (-5)\cdot(1\cdot7-2\cdot3)+9\cdot(4\cdot7-6\cdot3)-0\cdot(4\cdot2-6\cdot1)=85$$

DETERMINANT OF A MATRIX. The **determinant** obtained by considering the array of elements in a **square matrix** as elements of a determinant.

DETERMINISTIC GEOSTATISTICS. A deterministic interpretation of all procedures and results of the traditional **topoprobabilistic models**, including **kriging. Expected values** are replaced by spatial integrals and **regional characteristics. Probability distributions** are replaced by **experimental** distributions of values obtained by moving a **working neighborhood**. Deterministic geostatistics interprets the topoprobabilistic problem of **variogram** or **covariance** inference as the numerical approximation of a spatial integral, which leads to the development of **nonergodic experimental covariances**. The notion of deterministic geostatistics overlaps but is not synonymous with **transitive methods**.

DE WIJS FORMULA. A **model** of the **dispersion variance** $D^2(v|V)$ for the particular case when the **semivariogram** is of **de Wijsian semivariogram** type, and the **support** shape for the **field** V is the same as for the observations v

$$D^2(v|V) = \alpha \log_e(V|v)$$

The resulting **dispersion variance** is unbounded.

DE WIJSIAN SEMIVARIOGRAM. See **logarithmic semivariogram**.

DIAGONAL BIVARIATE DENSITY FUNCTION. A particular type of **probability density function** of two variables $f(\zeta_i, \zeta_j)$, which can be expanded in the form

$$f(\zeta_i,\zeta_j) = \sum_{k=0}^{\infty} \mathrm{E}[p_k(Z_i)\, q_k(Z_j)]\, p_k(\zeta_i)\, q_k(\zeta_j)\, f_1(\zeta_i)\, f_2(\zeta_j)$$

where $f_1(\zeta_i)$and $f_2(\zeta_j)$are **marginal probability density functions** of $f(\zeta_i, \zeta_j)$, and $p_k(\zeta_i)$ and $q_k(\zeta_j)$ are **orthogonal** square **integrable functions** with respect to $f_1(\zeta_i)$ and $f_2(\zeta_j)$, respectively. The functions $p_k(\zeta_i)$ and $q_k(\zeta_j)$ usually are polynomials. An example from **geostatistics** is the expansion of a **bivariate normal distribution** $g_r(x,y)$ with the **cross-correlation coefficient** r

$$g_r(x,y) = \sum_{n=0}^{\infty} \frac{r^n}{n!} H_n(x)\, H_n(y)\, g(x)\, g(y)$$

where $H_n(\cdot)$ is a **Hermite polynomial** and $g(\cdot)$ is the probability density function of a **normal distribution**.

DIAGONALIZED MATRIX. A **square matrix** whose **pivots** are all relative maxima, in the sense that each pivot is larger than any remaining **element of the matrix** along the **column of the matrix** and the **row of the matrix** comprising the pivot.

DIAGONAL MATRIX. A **square matrix** whose elements are all 0 except those along the **main diagonal**.

DILUTION RANDOM FUNCTION. Let $\Re^n$ be an n-dimensional **euclidean space** in which points $\boldsymbol{x}$ are generated according to a **Poisson process** $\boldsymbol{P}$ and used as the centers of spheres of random radius. Let $\boldsymbol{y}$ belong to $\Re^n$ and let $W_{\boldsymbol{x}}(\boldsymbol{y})$ be a **random function** defined inside the sphere centered at $\boldsymbol{x}$, and having **stochastic independence** from other random functions defined inside other spheres. The random function $Z(\boldsymbol{y})$ defined by

$$Z(\boldsymbol{y}) = \sum_{x \in \mathrm{P}} W_{\boldsymbol{x}}(\boldsymbol{y} - \boldsymbol{x})$$

is called the dilution random function and is used in **simulation** by the **random tokens method**.

DIRAC DELTA FUNCTION. The generalized **function**, $\delta(x)$, possessing the property that:

$$\int_{-\infty}^{\infty} \delta(x)\,dx = 1$$

$$\delta(x-a)=0, \quad \text{for } x \neq a$$

The delta function can be regarded as the limit of a **probability density function** as its **range** around a tends to 0.

DIRECT PROBLEM. Calculation of the response given by a mathematical **model** depending on a set of parameters; as opposed to the **inverse problem**.

DIRICHLET TESSELATION. A two-dimensional **Voronoi division**.

DISCRETE GAUSSIAN MODEL. Let v_i be a **partition** of a **field** into **supports** of equal size and shape. Let $z(v_i)$ be the **average value** of the **regionalized variable** $z(\boldsymbol{x})$ on the support v_i. Let $z(\overline{\boldsymbol{x}}_j)$ be the value of a point **sample** randomly and equiprobably located inside the support v_j. If we consider the regionalized variable $z(\boldsymbol{x})$ as a **realization** of the **stationary** type of **random function** $Z(\boldsymbol{x})$, then the discrete Gaussian model consists in assuming that the pair of **random variables** $Z(v_i)$ and $Z(\overline{\boldsymbol{x}}_j)$ follow a **bivariate normal distribution** for any i and j. The condition of random location of $\overline{\boldsymbol{x}}_j$ is required to assure the **consistency** of the discrete Gaussian model.

DISCRETE RANDOM VARIABLE. A **variate** that can assume values solely from a set without **accumulation points**, such as the set of integer numbers. Any variate that takes only a finite or a countable number of values is necessarily discrete. The **probabilities** associated with these points are provided by a **probability function**.

DISJUNCTIVE KRIGING. A nonlinear, **distribution**-dependent **estimator** of the form

$$Z^*(\boldsymbol{x}_0)=\sum_{i=1}^{n} f_i[Z(\boldsymbol{x}_i)]$$

where $Z^*(\boldsymbol{x}_0)$ is an estimator of the **regionalized variable** $z(\boldsymbol{x})$ at location $\boldsymbol{x}_0$ using n observations taken at sites $\boldsymbol{x}_i$, and $f_i[Z(\boldsymbol{x}_i)]$ is a **function** that must be square **integrable with respect to functions** of **univariate** distributions. Provided the **random field** is second order **stationary**, $f[Z(\boldsymbol{x}_i)]$ is found as the solution of a system of integral equations defined on the **conditional expectations**

$$\sum_{i=1}^{n} \mathrm{E}\left[f_i(Z(\boldsymbol{x}_i))\middle|Z(\boldsymbol{x}_j)\right]=\mathrm{E}\left[Z(\boldsymbol{x}_0)\middle|Z(\boldsymbol{x}_j)\right], \quad \text{for } j=1,2,\ldots,n$$

All the expectations can be computed directly from **bivariate** density function **models** $g[Z(\boldsymbol{x}_i),Z(\boldsymbol{x}_j)]$. Originally, disjunctive kriging was developed in terms of **bivariate normal distributions**, using **Hermite polynomials** to approximate the bivariate distributions. Disjunctive kriging is the most demanding of the **kriging** procedures in terms of **stationarity** conditions, computer resources, and mathematical understanding on the part of the user. Disjunctive kriging seems to provide better **estimates** of regionalized variables that cannot be modeled as **multinormal distributions**, but can be modeled by **isofactorial bivariate density functions**.

DISPERSION VARIANCE. A kind of **semivariogram** in which pairs of values defined over a small **support** are accepted in the **averaging** procedure as long as the pairs are contained inside a larger support. Let $Z_{v_1}(\boldsymbol{x})$ be the **random function** used to **model** the **regionalized variable** defined over the supports of size v_1 centered at x_i. In case of an **intrinsic** or **stationary** random function, the dispersion variance $D^2(v_1|v_2)$ is the **expected value** of the **mean** squared difference between the random function defined over those smaller supports of size v_1 and a random function $Z_{v_2}(\boldsymbol{x})$ defined over a **partition** set v_2 of the **field**

$$D^2(v_1|v_2) = E\left[\frac{1}{n}\sum_{i=1}^{n}\{Z_{v_1}(x_i) - Z_{v_2}(x_j)\}^2\right]$$

where x_j is the center for that support of size v_2 enclosing the supports of size v_1 centered at x_i. As a special case, v_2 can be the field itself. Suppose the sampling {4,9,8,6,5,4,7,3,2} resulting from assaying a continuous core of an **additive set function** into 1-inch cylinders. If we now consider 3-inch cylinders, the new values would be {7,5,4}. The dispersion variance of the 1-inch cylinders within the 3-inch cylinders is

$$\begin{aligned} D(1''\text{cyl.}|3''\text{cyl.}) &= \frac{1}{9}[(4-7)^2 + (9-7)^2 + (8-7)^2 + (6-5)^2 + (5-5)^2 + \\ &\quad + (4-5)^2 + (7-4)^2 + (3-4)^2 + (2-4)^2] \\ &= \frac{30}{9} \end{aligned}$$

Krige's formula is a generalization of the results to three sets of supports. It can be proved that the dispersion variance is equivalent to the difference of the two mean values of the semivariogram over the volumes v_1 and v_2

$$D^2(v_1|v_2) = \bar{\gamma}(v_2, v_2) - \bar{\gamma}(v_1, v_1) = F(v_2) - F(v_1)$$

where $F(\cdot)$ is the **F-function**.

DISTRIBUTION. A graphical or analytical representation showing the frequency of occurrence for **outcomes** of a **random variable**. Distributions of common use receive special names, such as **beta distribution, binomial distribution, exponential distribution, gamma distribution, lognormal distribution, negative binomial distribution, normal distribution, Poisson distribution, Sichel t distribution**, and the **three-parameter lognormal distribution**.

DISTRIBUTION FUNCTION. Given a real number x and a **random variable** X, the set $\{X \le x\}$ is an **event**. Its **probability** is a number depending on x; that is, it is a **function** of x . This function, denoted *F(x)*, is the distribution function and is defined as

$$F(x) = P\{X \le x\}$$

where the threshold x does not need to be a value of an **outcome**. *F(x)* fully characterizes the probabilistic description of all outcomes of X. If X is a **discrete random variable**, then

$$F(x) = \sum_{y_i \le x} P(X = y_i)$$

and if X is a *continuous random variable*, then

$$F(x) = \int_{-\infty}^{x} f(y)\,dy$$

whenever there is a **probability density function** $f(x)$. At all real numbers x at which the derivative exists

$$f(x) = \frac{d}{dx} F(x)$$

DOMAIN. The set of values over which a **function** is defined. The domain $\{\boldsymbol{x}:\boldsymbol{x} \in \Re^n\}$ of a **regionalized variable** is called a **field**.

DRIFT. The **expected value** of a **random function** $Z(\boldsymbol{x})$ at a point $\boldsymbol{x}$, $m(\boldsymbol{x}) = \mathrm{E}[Z(\boldsymbol{x})]$. The drift is a purely mathematical construct of a random function and has no physical reality. The concept of drift is usually linked to the physical idea of **trend**, which implies that the drift possesses a gentle and continuous spatial variation. Because of that constraint, drifts often are **modeled** by **basis functions**, usually low order polynomials $p^l(\boldsymbol{x})$ that, to further assure a simple surface, are defined within small **neighborhoods** relative to the size of the sampling **domain**.

$$m(\boldsymbol{x}) \approx \sum_{l=0}^{n} a_l p^l(\boldsymbol{x})$$

The **minimum mean square error estimator** for the drift is given by

$$m^*(\boldsymbol{x}) = \sum_{i=1}^{n} \lambda_i Z(\boldsymbol{x}_i)$$

The optimal value for the unknown λ_i in a **least squares method** sense is provided by $\boldsymbol{X} = \boldsymbol{A}^{-1}\boldsymbol{B}$ where

$$\boldsymbol{X} = \begin{bmatrix} \lambda_1 \\ \cdots \\ \lambda_i \\ \cdots \\ \lambda_n \\ \mu_0 \\ \mu_1 \\ \cdots \\ \mu_j \\ \cdots \\ \mu_m \end{bmatrix} \qquad \boldsymbol{B} = \begin{bmatrix} 0 \\ \cdots \\ 0 \\ \cdots \\ 0 \\ 1 \\ p^1(\boldsymbol{x}_0) \\ \cdots \\ p^j(\boldsymbol{x}_0) \\ \cdots \\ p^m(\boldsymbol{x}_0) \end{bmatrix}$$

$$\boldsymbol{A} = \begin{bmatrix} \mathrm{Cov}(0) & \cdots & \mathrm{Cov}(\boldsymbol{x}_1,\boldsymbol{x}_i) & \cdots & \mathrm{Cov}(\boldsymbol{x}_1,\boldsymbol{x}_n) & 1 & p^1(\boldsymbol{x}_1) & \cdots & p^j(\boldsymbol{x}_1) & \cdots & p^m(\boldsymbol{x}_1) \\ \cdot & \cdot & \cdot & \cdot & \cdot & \cdot & \cdot & \cdot & \cdot & \cdot & \cdot \\ \mathrm{Cov}(\boldsymbol{x}_i,\boldsymbol{x}_1) & \cdots & \mathrm{Cov}(0) & \cdots & \mathrm{Cov}(\boldsymbol{x}_i,\boldsymbol{x}_n) & 1 & p^1(\boldsymbol{x}_i) & \cdots & p^j(\boldsymbol{x}_i) & \cdots & p^m(\boldsymbol{x}_i) \\ \cdot & \cdot & \cdot & \cdot & \cdot & \cdot & \cdot & \cdot & \cdot & \cdot & \cdot \\ \mathrm{Cov}(\boldsymbol{x}_n,\boldsymbol{x}_1) & \cdots & \mathrm{Cov}(\boldsymbol{x}_n,\boldsymbol{x}_i) & \cdots & \mathrm{Cov}(0) & 1 & p^1(\boldsymbol{x}_n) & \cdots & p^j(\boldsymbol{x}_n) & \cdots & p^m(\boldsymbol{x}_n) \\ 1 & \cdots & 1 & \cdots & 1 & 0 & 0 & \cdots & 0 & \cdots & 0 \\ p^1(\boldsymbol{x}_1) & \cdots & p^1(\boldsymbol{x}_i) & \cdots & p^1(\boldsymbol{x}_n) & 0 & 0 & \cdots & 0 & \cdots & 0 \\ \cdot & \cdot & \cdot & \cdot & \cdot & \cdot & \cdot & \cdot & \cdot & \cdot & \cdot \\ p^j(\boldsymbol{x}_1) & \cdots & p^j(\boldsymbol{x}_i) & \cdots & p^j(\boldsymbol{x}_n) & 0 & 0 & \cdots & 0 & \cdots & 0 \\ \cdot & \cdot & \cdot & \cdot & \cdot & \cdot & \cdot & \cdot & \cdot & \cdot & \cdot \\ p^m(\boldsymbol{x}_1) & \cdots & p^m(\boldsymbol{x}_i) & \cdots & p^m(\boldsymbol{x}_n) & 0 & 0 & \cdots & 0 & \cdots & 0 \end{bmatrix}$$

and Cov(·) is the **covariance** of the original random function. If the covariance does not exist, optimal **estimation** of the drift is impossible and only the increments can be **estimated**.

DUAL KRIGING. Any **point kriging** system of equations has the property that the weights λ_i from the solution of the system of equations are a linear combination of the **structural functions** $K(\boldsymbol{x}_j - \boldsymbol{x}_0)$ and the **basis functions** $p^l(\boldsymbol{x}_0)$ on the right side of the system of equations

$$\lambda_i = \sum_{j=1}^{n} \alpha_j K(\boldsymbol{x}_j - \boldsymbol{x}_0) + \sum_{l=1}^{m} \beta_l\, p^l(\boldsymbol{x}_0)$$

The coefficients α_j and β_l are not **functions** of $\boldsymbol{x}_0$. The point **estimator** at location $\boldsymbol{x}_0$ has the same structure

$$z^*(\boldsymbol{x}_0) = \sum_{j=1}^{n} b_j K(\boldsymbol{x}_j - \boldsymbol{x}_0) + \sum_{l=1}^{m} c_l\, p^l(\boldsymbol{x}_0)$$

The coefficients b_j and c_l are obtained from the solution of the dual kriging system $\boldsymbol{X} = \boldsymbol{A}^{-1}\boldsymbol{B}$ where

$$\boldsymbol{X} = \begin{bmatrix} b_1 \\ \cdots \\ b_i \\ \cdots \\ b_n \\ c_0 \\ c_1 \\ \cdots \\ c_j \\ \cdots \\ c_m \end{bmatrix} \qquad \boldsymbol{B} = \begin{bmatrix} z(\boldsymbol{x}_1) \\ \cdots \\ z(\boldsymbol{x}_i) \\ \cdots \\ z(\boldsymbol{x}_n) \\ 0 \\ 0 \\ \cdots \\ 0 \\ \cdots \\ 0 \end{bmatrix}$$

$$\boldsymbol{A} = \begin{bmatrix}
\mathrm{K}(0) & \cdots & \mathrm{K}(\boldsymbol{x}_1,\boldsymbol{x}_i) & \cdots & \mathrm{K}(\boldsymbol{x}_1,\boldsymbol{x}_n) & 1 & p^1(\boldsymbol{x}_1) & \cdots & p^j(\boldsymbol{x}_1) & \cdots & p^m(\boldsymbol{x}_1) \\
\cdot & \cdot & \cdot & \cdot & \cdot & \cdot & \cdot & \cdot & \cdot & \cdot & \cdot \\
\mathrm{K}(\boldsymbol{x}_i,\boldsymbol{x}_1) & \cdots & \mathrm{K}(0) & \cdots & \mathrm{K}(\boldsymbol{x}_i,\boldsymbol{x}_n) & 1 & p^1(\boldsymbol{x}_i) & \cdots & p^j(\boldsymbol{x}_i) & \cdots & p^m(\boldsymbol{x}_i) \\
\cdot & \cdot & \cdot & \cdot & \cdot & \cdot & \cdot & \cdot & \cdot & \cdot & \cdot \\
\mathrm{K}(\boldsymbol{x}_n,\boldsymbol{x}_1) & \cdots & \mathrm{K}(\boldsymbol{x}_n,\boldsymbol{x}_i) & \cdots & \mathrm{K}(0) & 1 & p^1(\boldsymbol{x}_n) & \cdots & p^j(\boldsymbol{x}_n) & \cdots & p^m(\boldsymbol{x}_n) \\
1 & \cdots & 1 & \cdots & 1 & 0 & 0 & \cdots & 0 & \cdots & 0 \\
p^1(\boldsymbol{x}_1) & \cdots & p^1(\boldsymbol{x}_i) & \cdots & p^1(\boldsymbol{x}_n) & 0 & 0 & \cdots & 0 & \cdots & 0 \\
\cdot & \cdot & \cdot & \cdot & \cdot & \cdot & \cdot & \cdot & \cdot & \cdot & \cdot \\
p^j(\boldsymbol{x}_1) & \cdots & p^j(\boldsymbol{x}_i) & \cdots & p^j(\boldsymbol{x}_n) & 0 & 0 & \cdots & 0 & \cdots & 0 \\
\cdot & \cdot & \cdot & \cdot & \cdot & \cdot & \cdot & \cdot & \cdot & \cdot & \cdot \\
p^m(\boldsymbol{x}_1) & \cdots & p^m(\boldsymbol{x}_i) & \cdots & p^m(\boldsymbol{x}_n) & 0 & 0 & \cdots & 0 & \cdots & 0
\end{bmatrix}$$

For a given **neighborhood**, the dual system needs to be solved only once, in contrast to a direct **kriging** system that requires one inversion and as many **matrix** multiplications as there are locations to be **estimated**. However, with dual kriging it is not possible to determine the **estimation variance**.

EIGENVALUE. Each of the roots λ_i of the **characteristic equation of matrix *A***. The number of roots, n, is equal to the **order of the square matrix**; the roots are all real when the **matrix** is a **symmetric matrix**; and the sum of the roots is equal to the sum of the elements on the **main diagonal of the matrix**. Also, if $\boldsymbol{X}$ is a **vector** with as many rows as $\boldsymbol{A}$, then the eigenvalues are the solution to the equation $\boldsymbol{AX} = \lambda \boldsymbol{X}$ for the nontrivial case when all elements of $\boldsymbol{X}$ are not equal to 0. Geometrically, if the matrix is viewed as representing the coordinates of n points in an n-dimensional **euclidean space**, the eigenvalues represent half the length of the axes of an n-dimensional ellipsoid centered at the origin and passing through the n points. As an example, for the matrix

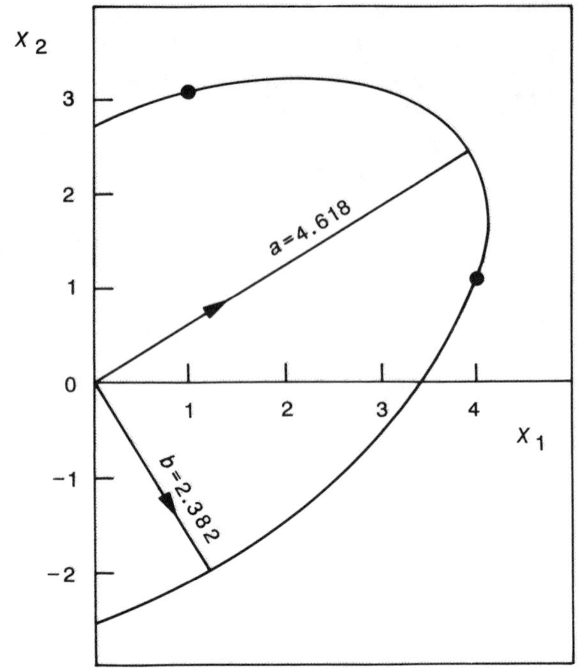

$$\begin{bmatrix} 4 & 1 \\ 1 & 3 \end{bmatrix}$$

the eigenvalues are $\lambda_1 = 2.382$ and $\lambda_2 = 4.618$. In the illustration, the vectors along the axes are the **eigenvectors**.

EIGENVALUE OF A MATRIX. See **eigenvalue**.

EIGENVECTOR. Let $\boldsymbol{A}$ be a **square matrix** and $\boldsymbol{X}$ be a **vector** of unknowns with as many rows as $\boldsymbol{A}$. Given the system of equations

$$\boldsymbol{AX} = \lambda_i \boldsymbol{X}$$

an eigenvector is a solution to the system of equations for an eigenvalue λ_i. Since there may be as many eigenvalues as the order n of **matrix *A***, there also will be a corresponding number of eigenvectors. Geometrically, if the matrix is viewed as representing the coordinates of n points in an n-dimensional **euclidean space**, then the eigenvectors can be regarded as vectors pointing in the direction of the axes of an n-dimensional ellipsoid centered at the origin and passing through those n points. As an example, for the matrix

$$\begin{bmatrix} 4 & 1 \\ 1 & 3 \end{bmatrix}$$

since the eigenvalues are $\lambda_1 = 2.382$ and $\lambda_2 = 4.618$, the corresponding two eigenvectors $\boldsymbol{X}_{\lambda_1}$ and $\boldsymbol{X}_{\lambda_2}$ are

$$\boldsymbol{X}_{\lambda_1} = \alpha_1[1.0 \quad (-1.618)] \qquad \boldsymbol{X}_{\lambda_2} = \alpha_2[1.0 \quad 0.618]$$

where α_i is an arbitrary constant. See the illustration at **eigenvalue**. The eigenvectors of a **covariance matrix** are called **principal components**.

ELEMENT OF A DETERMINANT. Each of the numbers or variables comprising the individual terms in a **determinant**.

ELEMENT OF A MATRIX. Each of the numbers or variables comprising the individual terms in a **matrix** array.

ELLIPTICAL ANISOTROPY. A synonym for **geometric anisotropy**.

ENSEMBLE. A synonym for **population**.

ENTROPY. In information **theory**, a **measure** of the **uncertainty** of knowledge. The more chaotic a system, the more unpredictable and higher is its entropy. The entropy, ε, of a **population** whose **probability density function** is $f(x)$, is defined as

$$\varepsilon = \mathrm{E}[-\log_e(f(X))]$$

Of all **continuous distributions** for which the probability density function exists and for which the second **moment** exists, the **normal distribution** has the largest entropy. Among all continuous distributions constrained by a fixed **range** of **outcomes**, the **uniform distribution** has the largest entropy. *Maximum entropy formalisms* seek to minimize any ordering of the system that does not stem from the **data**.

ERGODICITY. Generally, this term denotes a property of certain systems that develop through time according to probabilistic laws. Under certain circumstances, a system will tend in **probability** to a limiting form that is independent of the initial position from which it started. This is the *ergodic property*. A **stationary** stochastic process, $\{X(t,\omega): t \in T, \omega \in \Omega\}$, may be regarded as the set of all feasible **realizations** under the process. For a fixed value of Ω, ω_i, each realization $X(t,\omega_i)$ may have a **mean** m_i. If the process itself has a mean μ, the *ergodic theorem* of Birkhoff and Khintchine states that m_i exists for almost all realizations. If, in addition, $m_i = \mu$ for almost all realizations, the process is said to be ergodic. In this sense, ergodicity may be regarded as a form of the **law of large numbers** applied to stationary **stochastic processes**. Ergodicity implies that the **expected value**, a **model** parameter, can be approximated by the mean of an actual realization over a sufficiently large **domain**.

ERGODIC PROPERTY. See **ergodicity**.

ERGODIC THEOREM. See **ergodicity**.

ERROR. The deviation of a computed, observed, or measured quantity from the true, specified, or theoretically correct value of such quantity. The error is called positive or an *overestimation* if the result is greater than the true value, and negative or an *underestimation* if it is less than the true value. Sources of errors from particular identifiable causes such as truncation, rounding, cumulation, or surveying are easy to correct. In **geostatistics**, most errors are of unknown origin and are believed to be related to imperfect modeling, to parameter **estimation**, or to both causes.

ESTIMATE. A particular value of an **estimator** for a given sampling. For instance, the **sample** {7,11,5,9}, has a **mean** of 8 as estimated by the **arithmetic mean**.

ESTIMATION. The process of inferring an unknown value. (a) In traditional statistics, the term estimation is reserved for the evaluation of parameters of a **probabilistic** model, such as the estimation of a **range** in a **semivariogram**. (b) In **geostatistics**, the evaluation of a magnitude that the observer does not know, but which nevertheless exists independently of the observer and the amount of information at hand. It does not matter whether that magnitude is associated in a **model** to a **random variable** or is a parameter. In this sense, the term estimation is also used for **prediction** of the **realization** of a **random function** at an unsampled location.

ESTIMATION VARIANCE. The **variance** of the estimation **error**. A large estimation variance relative to the **estimate** implies that the estimate is poor. The **probability** is large that such an **estimation** is far from reality.

ESTIMATOR. A **random variable** resulting from the combination of the random variables involved in a sampling and used for **estimation**.

E-TYPE ESTIMATE. A **conditional expectation** form of **estimate** derived from a **conditional distribution** type of **model** provided by **multiple indicator kriging**. It is different from the estimate provided by direct **simple kriging** or **ordinary kriging**.

EUCLIDEAN SPACE. A **metric space** that is also a **linear space** and consists of all ordered sets $\{x_1, x_2, \ldots, x_i, \ldots, x_n\}$ of n real numbers. The distances $\|\boldsymbol{x} - \boldsymbol{y}\|$ between pairs of values are given by

$$\|\boldsymbol{x} - \boldsymbol{y}\| = \left[\sum_{i=1}^{n} (x_i - y_i)^2\right]^{\frac{1}{2}}$$

The number n is called the dimension of the space.

EVENT. A subset of a **sample space**. An event may comprise one or more **outcomes**, or none at all. The **probability** of an event is given by the **probability function**.

EVENT SPACE. A synonym for **sample space**.

EXACT INTERPOLATOR. An **interpolation** procedure that can consistently replicate **data** at the locations of observations. **Kriging** is an exact interpolator.

EXPECTED VALUE. Let X be a **random variable** having a **probability function** $f(x)$, and let $u(x)$ be a **function** of x . The expected value of $u(x)$ is denoted by the **operator** $\mathrm{E}[u(X)]$ and is defined, when it exists, by the integral

$$\int_{-\infty}^{\infty} u(x) f(x) dx$$

if X is a **continuous random variable**. If X is a **discrete random variable**, the expected value is defined, when it exists, by the summation

$$\sum_{x} u(x) f(x)$$

The expected value exists if and only if $\mathrm{E}\left[|u(X)|\right] < \infty$. For any given random variable and probability function, there are as many expected values as there are functions $u(x)$ that may exist. The concept can be generalized to more than one random variable, in which case it can be proved that the expected value of a **linear combination of variables** is equal to the linear combination of the expected values of these random variables X_i

$$\mathrm{E}\left[\sum_{i=1}^{n} a_i X_i\right] = \sum_{i=1}^{n} a_i \, \mathrm{E}[X_i]$$

for any coefficients a_i and random variables, whether the variables possess **stochastic dependence** or not. Some expected values are useful for the parametric characterization of the **distribution** and have special names, such as **covariance, drift, mean, semivariogram,** and **variance**. If $\{x_1, x_2, \ldots, x_i, \ldots, x_n\}$ is a **declustered** sampling of a **distribution**, the **arithmetic mean** of $u(x_i)$ is used as an **estimate** of the expected value.

EXPERIMENT. Any operation or process of doing or observing something happening under specified conditions that results in some final **outcome**.

EXPERIMENTAL COVARIANCE. An **estimate** of the **covariance** between two **random variables** $Z(\boldsymbol{x})$ and $Z(\boldsymbol{x}+\boldsymbol{h})$ given by

$$\mathrm{Cov}^*(\boldsymbol{h}) = \frac{1}{n(\boldsymbol{h})} \sum_{i=1}^{n(\boldsymbol{h})} z(\boldsymbol{x}_i+\boldsymbol{h})z(\boldsymbol{x}_i) - m^*$$

where $n(\boldsymbol{h})$ is the number of pairs of observations a distance $\boldsymbol{h}$ apart, and m^* is an estimate of the overall **mean** $m = \mathrm{E}[Z(x)]$. Unless the mean is correctly estimated so that $m^* \approx m$, which presupposes **ergodicity** of the **random function**, this traditional covariance estimate is **biased.** A better estimate is then given by the **nonergodic experimental covariance.**

EXPERIMENTAL SEMIVARIOGRAM. An **estimate** of the **semivariogram** based on sampling. For a **random field** $Z(\cdot)$ that is **intrinsic** or second order **stationary**, the estimate $\gamma^*(\boldsymbol{h})$ is

$$\gamma^*(\boldsymbol{h}) = \frac{1}{2n(\boldsymbol{h})} \sum_{i=1}^{n(\boldsymbol{h})} [z(\boldsymbol{x}_i+\boldsymbol{h}) - z(\boldsymbol{x}_i)]^2$$

where $n(\boldsymbol{h})$ is the number of pairs of observations a distance $\boldsymbol{h}$ apart. In practice, there is often enough randomness in the observation locations that some tolerance must be allowed both in direction and distance to have a significant number of pairs. In that situation, $\boldsymbol{h}$ is not necessarily a distance between pairs of locations, but is rather a distance for a class of observations. The experimental semivariogram is an approximation of the **regional semivariogram.**

EXPONENTIAL DISTRIBUTION. A **random variable** X has an exponential distribution if the **range** of X is the set of nonnegative real numbers and its **probability density function** $f(x)$ is given by

$$f(x) = \lambda e^{-\lambda x}, \quad x \geq 0$$

where λ is a positive number. The distribution is a special case of the **gamma distribution.**

EXPONENTIAL SEMIVARIOGRAM. The **semivariogram model**

$$\gamma(h) = C\left(1 - e^{-\frac{h}{a}}\right)$$

The model approaches the **sill** asymptotically. A practical definition of the **range** is the distance $3a$ where the **semivariogram** is 0.95 C. Geometrically, a tangent at the origin intersects the asymptote C at **lag** a.

EXTENSION VARIANCE. Let v_1 be a **support** for which an **average value** property, $Z(v_1)$, such as a concentration, is unknown and must be **estimated** by observations taken on a support having a different volume, v_2:

$$Z(v_i) = \frac{1}{v_i} \int_{v_i} Z(\boldsymbol{x})d\boldsymbol{x}$$

The extension variance, $\sigma_e^2(v_1,v_2)$, is the **estimation variance** that results from forecasting the average property in v_1 using the average value defined over v_2. The extension variance is given by

$$\sigma_e^2(v_1,v_2) = 2\bar{\gamma}(v_1,v_2) - \bar{\gamma}(v_1,v_1) - \bar{\gamma}(v_2,v_2)$$

where $\bar{\gamma}(v_i,v_j)$ is the **mean semivariogram** between volumes v_i and v_j.

EXTERNAL DRIFT METHOD. A procedure typically used to describe non**stationary** regionalized variables that have been **sampled** more than once by different methods. In this application, the external drift method may be an alternative to **cokriging**. The least **accurate** survey is used to **estimate** the **drift** required by **kriging,** and the more accurate survey is used to **predict** the **regionalized var-**

iable of interest either through **intrinsic kriging** or **universal kriging**.

EXTRAPOLATION. See **interpolation**.

FACTORABLE RANDOM FUNCTION. Let k be a nonnegative integer. Let $Z(\boldsymbol{x})$ be a **random function** defined over an n-dimensional **euclidean space** $\Re^n$. Let $g_i(\zeta_i)$ and $g_j(\zeta_j)$ be the **univariate probability density functions** for the **random variables** $Z(\boldsymbol{x}_i)$ and $Z(\boldsymbol{x}_j)$, and let $g(\zeta_i,\zeta_j)$ be their **bivariate** density function. Let $\{p_k(\zeta_i)\}$ and $\{q_k(\zeta_j)\}$ be two **complete sets of orthogonal polynomials** in $L_2[\Re^n, g(\zeta_i)]$ and $L_2[\Re^n, g(\zeta_j)]$, respectively; namely, they are square **integrable functions with respect to** $g_i(\zeta_i)$ and $g_j(\zeta_j)$. The random function $Z(\boldsymbol{x})$ is called a factorable random function if

$$\theta(\zeta_i,\zeta_j)=\frac{g(\zeta_i,\zeta_j)}{g(\zeta_i)\,g(\zeta_j)}$$

is in $L_2[\Re^n\,\Re^n, g(\zeta_i)\,g(\zeta_j)]$, and

$$h_m(\zeta_i)=\int_{\Re}\theta(\zeta_i,\zeta_j)\,g(\zeta_j)\,q_m(\zeta_j)\,d\zeta_j$$

$$h_m(\zeta_j)=\int_{\Re}\theta(\zeta_i,\zeta_j)\,g(\zeta_i)\,p_m(\zeta_i)\,d\zeta_i$$

are **polynomials of degrees up to** $\boldsymbol{m}$ in ζ_i and ζ_j, respectively. The bivariate density function $g(\zeta_i,\zeta_j)$ of a **factorable random process** can always be expressed in the form of a **diagonal bivariate density function**, but the converse is not true. Factorable random functions are a special case of **orthogonal random functions**.

FACTORABLE RANDOM PROCESS. A unidimensional **factorable random function**.

FACTOR ANALYSIS. An observed **multivariate** set of variables $\{z_1, z_2, \ldots, z_i, \ldots, z_k\}$ may be presumed to have arisen from **correlation** with underlying set of variables $\{y_1, y_2, \ldots, y_j, \ldots, y_p\}$ called factors, with $k<p$. Each z_i can be represented as a **linear combination of the variables** y_j together with a unique part s_i

$$z_i = s_i + \sum_{j=1}^{p} a_{ij} y_j$$

Factor analysis is a general name for techniques that **estimate** the various parameters in this **model**. The objective is to reduce the dimensionality of the problem by replacing the original variables with estimates of a smaller number of factors. Unless additional assumptions are introduced, the model is not unique. Several **estimation** procedures have been proposed, resulting in a number of related factor approaches; some of these are computationally complex.

FACTORIAL KRIGING. A method similar to **Fourier analysis** for decomposing the spatial variability of a **regionalized variable**. The **model** $Z(\boldsymbol{x})$ for the **random function** is decomposed into the sum of n random functions $Z_k(\boldsymbol{x})$.

$$Z(\boldsymbol{x}) = \sum_{k=1}^{n} Z_k(\boldsymbol{x})$$

The values of $Z_k(\boldsymbol{x})$ are assumed to be independent of one another. The **covariance** $\mathrm{Cov}_k(\boldsymbol{h})$ of each one of the components $Z_k(\boldsymbol{x})$ is presumed to be a **nested structure** of the covariance $\mathrm{Cov}(\boldsymbol{h})$ of the original model $Z(\boldsymbol{x})$

$$\text{Cov}(\boldsymbol{h}) = \sum_{k=1}^{n} \text{Cov}_k(\boldsymbol{h})$$

Just as the decomposition of a spectrum into **harmonics** is an arbitrary process, in the absence of additional information, the modeling of Cov($\boldsymbol{h}$) as n nested structures is also arbitrary. Factorial kriging is the **cokriging** process that **estimates** the sum of any number of $Z_k(\boldsymbol{x})$ components and **filters** out those components that are not retained.

FAST FOURIER TRANSFORM. Any of the digital computer algorithms for calculating the discrete **Fourier transform** more rapidly than by direct evaluation. Most methods are iterative and take advantage of mathematical symmetry and redundancy. The algorithms require that **data** be arranged on regular square **grids**.

F-FUNCTION. The **mean semivariogram** over rectangular shapes of **support** V, such as line segments, rectangles, or rectangular prisms. The F-function is denoted by $F(\cdot)$:

$$F(V) = \bar{\gamma}(V, V)$$

The F-function is also equal to the **dispersion variance** of a point support within a rectangular support V:

$$F(V) = \text{D}^2(0|V)$$

FIELD. The sampling **domain** of a **regionalized variable**.

FILTERING. An operation intended to modify the characteristics of a certain process. The modifications can be achieved by discriminating on the basis of properties such as frequency, wavelength, amplitude, or **range** of the **covariance**.

FORWARD PROBLEM. An alternative term for **direct problem**.

FOURIER ANALYSIS. The **theory** of representing **functions** of a variable t as a weighted sum of a series of sinusoidal functions. Provided that t varies between 0 and L

$$f(t) = \frac{a_0}{2} + \sum_{j=1}^{\infty} \left[a_j \cos\left(\frac{2\pi}{L} jt\right) + b_j \sin\left(\frac{2\pi}{L} jt\right) \right]$$

The unknown coefficients a_j and b_j can be calculated evaluating the **Fourier transforms** of the sinusoidal functions and making use of the fact that the sinusoidal functions are **orthogonal functions**.

FOURIER SYNTHESIS. See **Fourier transform**.

FOURIER TRANSFORM. A **transform** $g(\nu)$ is the Fourier transform of $f(t)$ if

$$g(\nu) = \int_{-\infty}^{\infty} f(t) e^{i\nu t} dt$$

Under suitable conditions for $g(\nu)$, it follows that

$$f(t) = \frac{1}{2\pi} \int_{-\infty}^{\infty} g(\nu) e^{-i\nu t} d\nu$$

Finding $g(\nu)$ from $f(t)$ is called **Fourier analysis** and finding $f(t)$ from $g(\nu)$ is called *Fourier synthesis*. $g(\nu)$ is the complex spectrum. Whenever $f(t)$ is real, the real part of $g(\nu)$ is the cosine transform and the imaginary part is the sine transform.

FRACTAL. Any subset of an n-dimensional **euclidean space** for which the **Hausdorff–Besicovitch dimension** D exceeds the dimension of the space. The principal characteristic of fractals is that they keep the same appearance despite changes in scale. For example, a branch of a tree has the same geometric form as the entire tree, and a mound looks like a small-scale mountain. Not all fractals occur naturally. Some can be generated artificially following pseudorandom or deterministic rules, such as a sequence that can be continued ad infinitum, as suggested in the illustration. The perimeter of the resulting fractal made from unidimensional line segments has $D = \log 4 / \log 3 \approx 1.2618$. In **geostatistics**, fractals have been used in connection with **simulation** of **random functions** with **stationary** increments and no finite **variance**, such as those random functions characterized by the **power semivariogram model**.

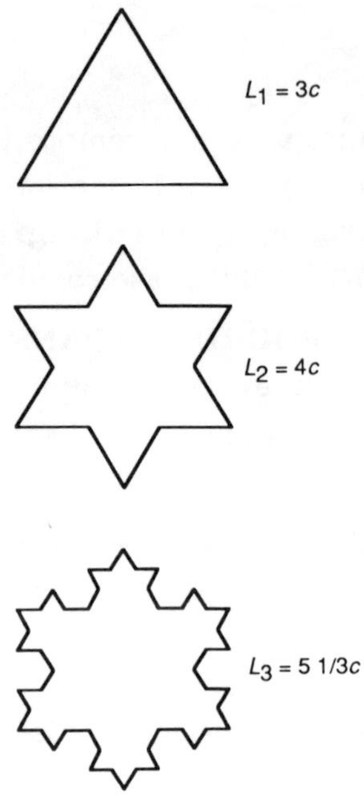

FRACTAL DIMENSION. An alternative term for **Hausdorff–Besikovitch dimension**.

FRACTILE. A synonym for **quantile**.

FREQUENCY CURVE. An alternative term for **frequency diagram**.

FREQUENCY DIAGRAM. A graphic representation of the relative proportion of items in a given number of categories in a **sample**. See also **histogram**.

FREQUENCY FUNCTION. Any nonnegative **function**, $f(x)$, satisfying

$$\int_{-\infty}^{\infty} f(x)dx = 1$$

FUNCTION. A rule associating exactly one element from one set, the **domain**, with one element of another set, the **range**. For instance, the function $f(x) = 1 + 2x^2$ turns 1 into 3, 2 into 9, 2.5 into 13.5, and so on.

FUZZY INDICATOR DATA. A synonym for **preposterior distribution** values at specific locations where supplementary information about the **regionalized variable** exists. It is not equivalent to the **stationary** type of **prior distribution** for a **random function**.

GAMMA DISTRIBUTION. Let X be a **random variable** on positive real numbers and let α and β be two positive numbers. The random variable X follows a gamma distribution if its **probability density function** is given by

$$f(x) = \frac{1}{\Gamma(\alpha)\beta^{\alpha}} x^{\alpha-1} e^{-\frac{x}{\beta}}, \quad x > 0$$

where $\Gamma(\cdot)$ is the **gamma function**. The **mean** is $\alpha\beta$ and the **variance** is $\alpha\beta^2$. The distribution is called an **exponential distribution** when α is equal to 1.

GAMMA FUNCTION. Let x be a positive number. The gamma function $\Gamma(x)$ is the integral

$$\Gamma(x) = \int_0^\infty t^{x-1} e^{-t}\, dt$$

It can be shown that $\Gamma(x)=1$; $\Gamma(x+1)=x\Gamma(x)$; and $\Gamma(n)=(n-1)!$, when n is any positive integer greater than 1.

GAUSSIAN ANAMORPHOSIS. A **bijection** between two **distribution functions** in which one of the distribution functions is a **standard normal distribution**. If $F(x)$ is the distribution function for the original distribution and $G(y)$ is the standard normal distribution, then the Gaussian anamorphosis is denoted by $y=G^{-1}[F(x)]$ and the inverse Gaussian anamorphosis is written as $x=F^{-1}[G(y)]$. This bijection is also known as the *normal score transform*. In the presence of discontinuities in $F(x)$, special provisions must be taken to ensure the one-to-one correspondence.

GAUSSIAN DISTRIBUTION. An alternative term for **normal distribution**.

GAUSSIAN SEMIVARIOGRAM. The **semivariogram model**

$$\gamma(h) = C\left[1 - e^{-\left(\frac{h}{a}\right)^2}\right]$$

The **sill** is approached asymptotically. A practical rule is to consider as the **range** the distance $\sqrt{3}\,a$, for which the **semivariogram** is 0.95 C. The graph of the model has **parabolic behavior** near the origin.

GENERALIZED COVARIANCE OF ORDER ν. Any of the **functions** $K(\cdot)$ of an **intrinsic random function of order** ν, $Z(x)$, that are used to calculate the **variance** of **permissible combinations of order** ν

$$\mathrm{Var}\left[\sum_{i=1}^{m} \alpha_i\, Z(\boldsymbol{x}_i)\right] = \sum_{i=1}^{m} \sum_{j=1}^{m} \alpha_i\, \alpha_j\, K(\boldsymbol{x}_i - \boldsymbol{x}_j)$$

Any **permissible coefficients of order** ν are shown by α_i. Generalized covariances are required by the **theory** behind **intrinsic kriging**. For a given intrinsic random function of order ν, there is an infinite number of solutions $K(\cdot)$. Any difference between a pair of solutions $K(\cdot)$ is an even polynomial of degree no larger than 2ν. Direct inference of such a covariance is not possible because of the implicit form of the equation. For that reason, generalized covariances are **modeled** using a parametric family of polynomial covariances for which the parameters are determined by a type of **cross-validation**.

GENERALIZED INCREMENTS OF ORDER ν. In the **theory** of finite differences, the generalized increment of order ν at location x of a **function** $f(x)$ is the value

$$\sum_{i=0}^{\nu+1} (-1)^i \begin{bmatrix} \nu+1 \\ i \end{bmatrix} f[x + a(\nu+1-i)]$$

where a is a scaling factor. Generalized increments of order ν are identical to finite differences of order $\nu+1$ and a special case of **permissible combinations of order** ν**–1**. Generalized increments of order ν **filter** out powers up to degree ν at the location x

$$\sum_{i=0}^{\nu+1} (-1)^i \begin{bmatrix} \nu+1 \\ i \end{bmatrix} [x + a(\nu+1-i)]^k = 0, \quad \text{for all } k \le \nu$$

GENERALIZED SEMIVARIOGRAM. For an **intrinsic random function of order** ν, $Z(\boldsymbol{x})$, the generalized semivariogram, $\gamma_\nu(\boldsymbol{h})$, is

$$\gamma_\nu(\boldsymbol{h}) = \begin{bmatrix} 2+2\nu \\ 1+\nu \end{bmatrix}^{-1} \mathrm{Var}[I_\nu(\boldsymbol{h})]$$

where $I_\nu(\boldsymbol{h})$ are **stationary** increments of order ν given by

$$I_\nu(\boldsymbol{h}) = \sum_{i=0}^{\nu+1} (-1)^i \begin{bmatrix} 1+\nu \\ i \end{bmatrix} Z[\boldsymbol{x}_j + (1+\nu-i)\boldsymbol{h}]$$

and $\boldsymbol{h}$ is a multiple of the sampling interval. The **generalized covariance** $K_\nu(\boldsymbol{h})$ and the generalized semivariogram $\gamma_\nu(\boldsymbol{h})$ of the same order ν are related by

$$\gamma_\nu(\boldsymbol{h}) = \begin{bmatrix} 2+2\nu \\ 1+\nu \end{bmatrix}^{-1} \sum_{i=-\nu-1}^{\nu+1} (-1)^i \begin{bmatrix} 2+2\nu \\ 1+\nu+i \end{bmatrix} K_\nu(i\,\boldsymbol{h})$$

In general, $K_\nu(\boldsymbol{h})$ cannot be expressed as a linear **function** of $\gamma_\nu(\boldsymbol{h})$. The exception is for $\nu = 0$, in which case $K_0(\boldsymbol{h}) = -\gamma_0(\boldsymbol{h}) = -\gamma(\boldsymbol{h})$. In practice, to calculate a generalized semivariogram, the **regionalized variable** must be **sampled** at regular intervals along a line and $\boldsymbol{h}$ is restricted to values that are multiples of the sampling interval.

GENERAL RELATIVE SEMIVARIOGRAM. A **relative semivariogram** over the entire field. The most common **estimate** has the same form as the **local relative semivariogram** with $k=1$. The number of pairs available is higher than those for any true local relative semivariogram with $k>1$, but the **intrinsic hypothesis** is more difficult to satisfy.

GEOMETRIC ANISOTROPY. The characteristic of a **regionalized variable** of possessing a spatial continuity that is directionally dependent, but which can be made isotropic by a linear **transformation** of the original coordinate system, as it is the case of a **semivariogram** with constant **sill**, but varying **range**. The correction is one that transforms any ellipse or ellipsoid into a circle or sphere. It consists of a sequence of rotations to align the coordinate axes with the main axes of the ellipsoid, then an **affinity transformation** that shrinks or expands some of these coordinate axes. Limiting cases, in which an ellipse is reduced to a line segment, or an ellipsoid is reduced to an ellipse, correspond to a special type of **zonal anisotropy** where spatial continuity is made independent of one or up to $(n-1)$ orthogonal directions in the **euclidean space** $\Re^n$.

GEOMETRIC AVERAGE. A synonym for **geometric mean**.

GEOMETRIC COVARIOGRAM. Let $k(x)$ be an **indicator** of a set according to definition (a). The geometric covariogram $K(h)$ is the **function**

$$K(h) = \int_S k(x)k(x+h)\,dx$$

The geometric covariogram is a measurement of the area of intersection of the set S with its translation by an amount h.

GEOMETRIC FIELD. A synonym for **support**, definition (b).

GEOMETRIC MEAN. The geometric mean $\bar{x}_g$ is the positive nth root of the product of n positive numbers

$$\bar{x}_g = \sqrt[n]{\prod_{i=1}^{n} x_i}$$

For any set in which not all of the numbers are identical, the geometric mean is always larger than the **harmonic mean** but smaller than the **arithmetic mean**.

GEOMETRIC SUPPORT. An alternative term for **support**, definition (a).

GEOSTATISTICS. The term has been used with two conflicting connotations. (a) The study of phenomena that fluctuate in space. Originally the field started with the objective of improving forecasts of ore grade and reserves in mines, but the mathematical generality of the approach led to the application of geostatistics to other problems in the earth sciences. Although not all geostatistical methods are **probabilistic** in nature, the most important accomplishments have been in **estimation** and forecasting, extending the probabilistic methods of **stochastic processes** and **time series** analysis to the spatial **domain**. This concept of geostatistics appeared in the early 1960s and is widely accepted outside North America, especially in South Africa and France, where early developments took place. The exact extent of geostatistics remains loosely defined, and its boundaries overlap related topics such as mathematical morphology, spatial statistics, image analysis, and remote sensing. This is the sense in which geostatistics has been used in the preparation of this glossary. (b) The application of statistical methods or the collection of statistical **data** for use in the earth sciences, particularly in geology. This usage is accepted primarily in the United States and was introduced almost contemporaneously at a time when quantitative methods in geology received an important boost from widespread availability of computers. In this sense, geostatistics is part of the wider discipline of mathematical or quantitative geology.

GLOBAL. Of, relating to, or involving the entire **field** of a **regionalized variable**.

GLOBAL ESTIMATION. An **estimation** that is based on the whole **field** of the **regionalized variable** under study, and all the available **data**. In **linear geostatistics**, global estimation consists of calculating the **variance** of an **estimator** that has an a priori given expression. This requires the use of approximation formulas. In **nonlinear geostatistics**, global estimation has been mainly devoted to the forecasting of **selectivity curves**.

GRADED SEMIVARIOGRAM. The **semivariogram** of **regionalized variables** which have been subjected to the process of **grading**.

GRADING. A special type of **regularization** in an n-dimensional **field**, which provides the **average value** over an interval L in a fixed direction. The coordinate axes are oriented so that the direction of grading coincides with axis x_n; the grading of a **regionalized variable** is then

$$Z_G(x_1,x_2,\ldots,x_{n-1}) = \frac{1}{L}\int_L Z(x_1,x_2,\ldots,x_{n-1},x_n)\,dx_n$$

The resulting regionalized variable, $Z_G(x_1,x_2,\ldots,x_{n-1})$, is called the grading of a point regionalized variable over a constant thickness L.

GRID. A rectangular reference mesh on a plane formed by two sets of regularly spaced orthogonal lines. Grids are used to determine the location of observations in regular and **stratified random sampling** as well as the location of **estimates** of a **regionalized variable**. Most automatic **contouring** procedures require that the set of values to be displayed be arranged at the nodes of a closely spaced grid. **Conditional simulation** and **kriging** are used, among other applications, to create a grid of estimates when the original **data** are irregularly spaced, too sparse, or both.

HARD DATA. An informal term for reliable sampling consisting of observations measured with a high degree of **accuracy**. Hard data allow definition of all of the components of **indicator data columns**.

HARMONIC. Each of the sinusoidal frequency components of a wave form.

HARMONIC MEAN. The harmonic mean, $\bar{x}_h$, is the reciprocal of the **arithmetic mean** of the reciprocals of n positive numbers,

$$\bar{x}_h = \frac{n}{\sum_{i=1}^{n} \frac{1}{x_i}}$$

For a set in which not all the numbers are identical, the harmonic mean is always smaller than the **arithmetic mean** and the **geometric mean.**

HAUSDORFF–BESICOVITCH DIMENSION. Let A be a subset in an n-dimensional **euclidean space,** and let $S_\alpha(A)$ be the **Hausdorff measure** of the subset. The Hausdorff–Besicovitch dimension is a unique positive real number D such that $S_\alpha(A)$ is infinite for $\alpha < D$ and vanishes for $\alpha > D$. The integer part of the Hausdorff–Besicovitch dimension is equal to the dimension n of the subset, and the remainder is a **measure** of tortuosity. $S_\alpha(A)$ is used to characterize **fractals.** There are several methods used to **estimate** the Hausdorff–Besicovitch dimension. The simplest of these is a graphical procedure used to determine the Hausdorff–Besicovitch dimension of an irregular line such as a coastline. The procedure is based on the fact that the length of the irregular line depends on the length of the ruler used for measuring. Let L_i be the total length measured using a ruler of length r_i in a survey using at least two rulers of different sizes. The slope of the **regression line** of $\log(L)$ as a **function** of $\log(r)$ is equal to $(1-D)$.

HAUSDORFF MEASURE. Let α be any real number larger than 0, let A be a subset of some n-dimensional **euclidean space,** and let B_i be a solid n-dimensional sphere of radius d_i smaller than ε. If σ is a countable collection of such spheres whose union covers A, then the Hausdorff measure, $S_\alpha(A)$, is defined as

$$S_\alpha(A) = \lim_{\varepsilon \to 0} \mathrm{I}_\sigma\Big(\sum_{i=1}^{m_\sigma} d_i^\alpha\Big)$$

where $\mathrm{I}_\sigma(\cdot)$ is the **infimum** taken over all possible collections of spheres. As α increases, the measure decreases. It can be shown that $S_\alpha(A)$ can take at most three values: 0, ∞, and a single nonzero finite value D that is known as the **Hausdorff-Besicovitch dimension** and is used in the characterization of **fractals.**

HEAVISIDE'S FUNCTION. The special **step function** $H(r)$ defined by

$$H(r) = \begin{cases} 0, & \text{if } r = 0 \\ 1, & \text{if } r > 0 \end{cases}$$

HERMITE MODEL. Let $\boldsymbol{h}$ be the distance between two **random variables** X and Y of a **random function** that is **stationary.** Let $R(\boldsymbol{h})$ be a **correlogram** of the two random variables. Let $H_n(\cdot)$ be a **Hermite polynomial,** and let $g(\cdot)$ be the **probability density function** of a **normal distribution.** The Hermite model is the **isofactorial bivariate density function,** $f(x,y)$, given by

$$f(x,y) = \sum_{n=0}^{\infty} \frac{\rho_n(\boldsymbol{h})}{n!} H_n(x)\, H_n(y)\, g(x)\, g(y)$$

where $\rho_n(\boldsymbol{h})$ is the **expected value,** $\mathrm{E}[R^n(\boldsymbol{h})]$. The **bigaussian model** and **mosaic model** are particular cases of Hermite models.

HERMITE POLYNOMIALS. The class of **functions** $H_n(x)$ obtained by finding the **orthogonal functions** over the interval $(-\infty,\infty)$ with respect to the weight function

$$w(x) = \frac{1}{\sqrt{2\pi}} e^{-\frac{x^2}{2}}$$

Hermite polynomials can be generated by the **recursion** formula

$$H_{n+1}(x) + x H_n(x) + n H_{n-1}(x) = 0, \quad n \geq 1$$

or the corresponding **Rodrigues formula**

$$H_n(x) = \frac{1}{w(x)} \frac{d^n w(x)}{d x^n}$$

The first three Hermite polynomials are

$$H_0(x) = 1$$
$$H_1(x) = -x$$
$$H_2(x) = x^2 - 2$$

It can be shown that

$$\int_{-\infty}^{\infty} H_n(x) H_m(x) w(x) dx = n! \, \delta_n^m$$

where δ_n^m is the **Kronecker delta**. Hermite polynomials are used in **disjunctive kriging** for the approximation of **distribution functions**.

HERMITIAN MATRIX. A **square matrix** such that every **element of the matrix** a_{ij} is equal to the **conjugate complex number** of a_{ji}

$$a_{ij} = \overline{a}_{ji}$$

A Hermitian matrix with real elements is a **symmetric matrix**.

HETEROSCEDASTIC. The condition of not being **homoscedastic**.

HETEROTOPIC. A **sampling** is heterotopic if one or more of the **regionalized variables** in a **coregionalization** are not consistently **sampled** at all locations.

HIGH-GRADE DESTRUCTURATION. The condition where raising an arbitrary threshold value z_0 lowers the spatial **correlation** for those **realizations** of a **regionalized variable** above z_0. This phenomenon is observed in a mineralized deposit that is a mixture of a low-to-**median** grade **structured** mineralization with a high-grade random mineralization, such as gold nuggets. **Models** based on **isofactorial bivariate density functions** allow destructuration of high grades, but also impose a symmetric destructuration of low grades that is not necessarily observed.

HIGH-GRADE STRUCTURATION. The condition where raising an arbitrary threshold value z_0 increases the spatial **correlation** for those **realizations** of a **regionalized variable** above z_0. The correlations are usually strong and **anisotropic**. This phenomenon commonly is found in precious metal deposits where a high-grade, highly **structured** vein-type mineralization is intermingled with a low-grade, poorly structured background. **Models** based on **isofactorial bivariate density functions**, in particular **multinormal distributions**, cannot be used to represent such phenomena because they imply a symmetric **high-grade destructuration** of very low and very high grades. A more satisfactory approach is provided by **multiple indicator kriging**, which permits different **semivariogram models** for different threshold values.

HILBERT SPACE. A **vector space** whose axioms relate to addition, multiplication, and convergence. In the formal language of mathematics, a Hilbert space is an **inner product space** that is a **complete space** with respect to the **norm**

$$\|\boldsymbol{x}\| = \sqrt{\langle \boldsymbol{x}, \boldsymbol{x} \rangle}$$

where $\langle \boldsymbol{x}, \boldsymbol{x} \rangle$ denotes the **inner product.**

HISTOGRAM. A **frequency diagram** constructed by dividing the **range** of a variable into successive intervals, not necessarily of the same lengths, and drawing rectangles with the intervals as bases and areas proportional to the frequencies for the classes described by the intervals. In the most common display, all intervals are the same length and the heights of the rectangles are proportional to the frequencies. When the range is divided into **quantiles**, all rectangles are the same height and the interval lengths become proportional to the frequencies.

HOLE EFFECT. A condition in which a **structural function** is not monotonically increasing. A hole effect does not necessarily preclude the presence of a **sill** and may reflect **nested structures** or periodicities in the **random field.**

HOLE EFFECT SEMIVARIOGRAM MODEL. Any of the **functions** suitable for **modeling** a **hole effect**, such as

$$\gamma(h) = C\left(1 - \frac{\sin(ah)}{ah}\right)$$

HOMOGENEITY. (a) A property of a **Markov process** denoting invariance under translation of **transition probabilities.** Strictly **stationary** Markov processes are homogeneous, but the converse is not necessarily true. (b) A qualitative characteristic denoting that physical properties between elemental volumes of a material have the same value regardless of their locations. (c) The property of two or more **populations** of sharing the same **distribution function.** By extension the term also is applied to **samples** of such populations.

HOMOSCEDASTIC. Let $\{X_{i1}, X_{i2}, \ldots, X_{ij}, \ldots, X_{in}\}$ be observations from several sources that are independent and identically distributed. (a) In the broadest sense, if the **variances** of each X_{ij} are all equal, then the **ensemble** is homoscedastic. (b) In the narrower sense used in **geostatistics**, if any dispersion characteristic, such as the **estimation variance**, is independent of the numerical value of the observations, that variance is said to be homoscedastic.

h-SCATTERGRAM. A **scattergram** of all pairs of **realizations** of $Z(\boldsymbol{x})$ and $Z(\boldsymbol{x}+\boldsymbol{h})$ a distance $\boldsymbol{h}$ apart. The moment of inertia of all points around the main bisector is equal to the **experimental semivariogram** for **lag** $\boldsymbol{h}$. The h-scattergram is helpful when looking for anomalous pairs in a particular lag of the experimental semivariogram.

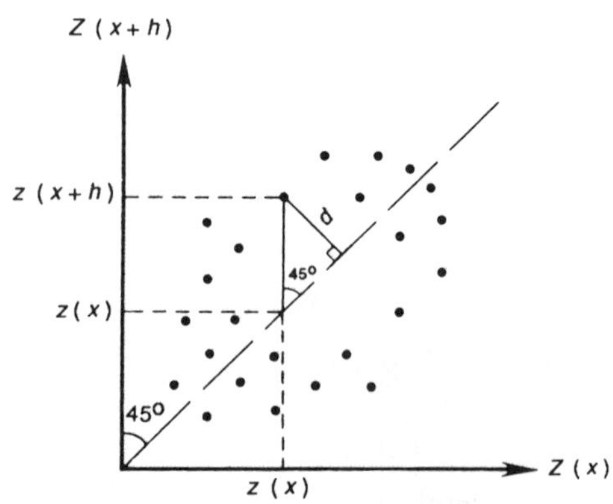

Copyright source: Hohn, 1988

IDENTITY MATRIX. A **square matrix** in which all elements on the **main diagonal of the matrix** are 1 and all off-diagonal elements are 0, such as

$$I = \begin{bmatrix} 1 & 0 & 0 \\ 0 & 1 & 0 \\ 0 & 0 & 1 \end{bmatrix}$$

IMPULSE. A synonym for **Dirac delta function.**

INDICATOR. (a) Let S be a subset of U and let $x \in U$. The indicator $k(x)$ of S is defined by

$$k(x) = \begin{cases} 0, & \text{if } x \notin S \\ 1, & \text{if } x \in S \end{cases}$$

(b) For a given **regionalized variable**, a **transformation** of observations $z(\boldsymbol{x})$ into binary numbers depending on whether the observations are above or below a threshold value z:

$$i(\boldsymbol{x}, z) = \begin{cases} 0, & \text{if } z(\boldsymbol{x}) > z \\ 1, & \text{if } z(\boldsymbol{x}) \le z \end{cases}$$

The indicator is another **random function**. For a **univariate** type of **distribution**, it can be shown that

$$\mathrm{E}[I(\boldsymbol{x}, z)] = F(z)$$

where $F(\cdot)$ is the **cumulative distribution function** for a **stationary** random function. In addition,

$$\mathrm{E}[I(\boldsymbol{x}, z_1)\mathrm{I}(\boldsymbol{x}+\boldsymbol{h}, z_2)] = F(\boldsymbol{h}; z_1, z_2)$$

where $\mathrm{E}[I(\boldsymbol{x}, z_1)I(\boldsymbol{x}+\boldsymbol{h}, z_2)]$ is the noncentered **indicator cross-covariance** and $F(\boldsymbol{h}; z_1, z_2)$ denotes the **bivariate** cumulative distribution for the pair of **random variables** $Z(\boldsymbol{x})$ and $Z(\boldsymbol{x}+\boldsymbol{h})$.

INDICATOR COKRIGING. **Cokriging** of the **indicator data columns** that result from considering more than one threshold at a time. **Probability kriging** and **indicator principal components kriging** are approximations to indicator cokriging.

INDICATOR COVARIANCE. Given a **random function** $Z(\boldsymbol{x})$ and a threshold z, the **covariance** for the **indicator,** $I(\boldsymbol{x}, z)$. The noncentered indicator covariance $\mathrm{E}[I(\boldsymbol{x}, z)I(\boldsymbol{x}+\boldsymbol{h}, z)]$ is a two-point **measure** of spatial connectivity that quantifies the **probability** that two **random variables** $Z(\boldsymbol{x})$ and $Z(\boldsymbol{x}+\boldsymbol{h})$ do not jointly exceed a given threshold z

$$\mathrm{E}[I(\boldsymbol{x}, z)I(\boldsymbol{x}+\boldsymbol{h}, z)] = P(Z(x) \le z, Z(\boldsymbol{x}+\boldsymbol{h}) \le z)$$

INDICATOR CROSS-COVARIANCE. A **covariance** between **indicators** associated with different thresholds. This cross-covariance is related to the **bivariate distribution** of the continuous **random function** from which the indicators are derived.

INDICATOR DATA COLUMN. Given information about a **regionalized variable** at location $\boldsymbol{x}$ and a set of thresholds $\{z_1, z_2, \ldots, z_i, \ldots, z_n\}$, the indicator data column is the **vector** of corresponding **indicator** transforms $\{i(\boldsymbol{x}, z_1), i(\boldsymbol{x}, z_2), \ldots, i(\boldsymbol{x}, z_i), \ldots, i(\boldsymbol{x}, z_n)\}^T$. When this information consists of **hard data,** all n components in the vector exist, but when this information consists of **soft data** some of the indicator **transforms** may not be defined, and the number of components may be less than n.

INDICATOR KRIGING. **Simple kriging** or **ordinary kriging** applied to **indicator** data sharing the same threshold. The indicator **data** and **covariances** come directly from the **sampling**, not from any prior **model**. Indicator kriging aims at providing a model for the **conditional distribution** of the unknown attribute at any unsampled location. The approach is simpler than **indicator co-**

kriging, but does not allow capturing the cross-information from indicator data having different thresholds.

INDICATOR PRINCIPAL COMPONENT KRIGING. **Kriging** of the **principal components** of the **covariance matrix** of the n **indicators** $I(\boldsymbol{x}, z_i)$, corresponding to n different thresholds. After an inverse **transformation**, this form of kriging provides an approximation to **indicator cokriging** and a **model** for **conditional distributions**.

INDIRECT LOGNORMAL CORRECTION. A modified version of the **lognormal correction** designed to adjust the change in the **mean** that results when applying the lognormal correction to a **distribution** that is not a **lognormal distribution**. The correction is done by rescaling the **quantiles** $q_l(p)$ of a lognormal correction by the ratio of the original mean, m_s, over the mean m_l of the values after the lognormal correction

$$q_i(p) = \frac{m_s}{m_l} q_l(p)$$

where $q_i(p)$ is a quantile of the new distribution obtained by applying the indirect lognormal correction. In addition to rescaling, the indirect lognormal correction lowers the **skewness** and increases the symmetry of the distribution.

INFIMUM. The greatest lower bound of an ordered set.

INFINITE DIVISIBILITY. The characteristic of a **random variable** X expressing that, for any number n, it is possible to find a collection of random variables X_i that are mutually **stochastically independent**, have the same **distribution function**, and are such that:

$$X = \sum_{i=1}^{n} X_i$$

This property is sometimes useful in simplifying some formulations, such as stochastic **simulation** for instance.

INFINITE DIVISIBILITY OF A UNION. A property of a **random set** S that is **closed** and such that, for any number n, it is possible to find a collection of closed random sets, S_i, that are mutually **stochastically independent**. In addition the sets have the same **Choquet's capacity** and satisfy

$$S = \bigcup_{i=1}^{n} S_i$$

INFORMATION EFFECT. If a **regionalized variable** may be **sampled** progressively, then future **estimations** may be based on **data** not available for early estimations. The impact of this additional information on future studies is referred to as the information effect.

INNER PRODUCT. In a complex **linear space** V, the complex valued **function** $\langle \boldsymbol{x}, \boldsymbol{y} \rangle$ whose **domain** is the set of all ordered pairs V whose **range** is the set of complex numbers satisfying the following axioms:

i. $\langle \boldsymbol{x}, \boldsymbol{y} \rangle = \langle \overline{\boldsymbol{x}, \boldsymbol{y}} \rangle$

ii. $\langle \lambda \boldsymbol{x}, \boldsymbol{y} \rangle = \bar{\lambda} \langle \boldsymbol{x}, \boldsymbol{y} \rangle$

iii. $\langle \boldsymbol{x} + \boldsymbol{y}, \boldsymbol{z} \rangle = \langle \boldsymbol{x}, \boldsymbol{z} \rangle + \langle \boldsymbol{y}, \boldsymbol{z} \rangle$

iv. $\langle \boldsymbol{x}, \boldsymbol{x} \rangle \begin{cases} = 0, & \text{if and only if } \boldsymbol{x} = 0 \\ > 0, & \text{otherwise} \end{cases}$

where the overbars indicate **conjugate complex numbers**. In a two-dimensional **euclidean space**, the inner product $\langle \eta, \xi \rangle$ of two **vectors** η and ξ is equal to the product of their length and the cosine of the angle θ between them,

$$\langle \eta, \xi \rangle = |\eta||\xi| \cos(\theta)$$

INNER PRODUCT SPACE. A **linear space** with an **inner product**.

INTEGRABLE FUNCTION. A **function**, *f(x)*, whose integral exists in some specified sense and is finite

$$\int_{-\infty}^{\infty} |f(x)| dx < \infty$$

INTEGRABLE FUNCTION WITH RESPECT TO A SECOND FUNCTION. The **function** *f(x)* is *k*th integrable with respect to a second function *g(x)* if the *k*th power of *f(x)* weighted by *g(x)*

$$f^k(x)\, g(x)$$

is an **integrable function**. If $\Re$ is the **domain** of integration, *f(x)* is said to be a function of the **Hilbert space**, $L_2[\Re, g(x)]$.

INTEGRAL RANGE. Let Cov($\boldsymbol{h}$) be the **covariance** of a **second order random field**. Its integral range A is defined by

$$A = \frac{1}{\text{Cov}(0)} \int_{\Re^n} \text{Cov}(\boldsymbol{h}) d\boldsymbol{h}$$

The integration is calculated over the entire n-dimensional **euclidean space**. The dimensions of A are those of length, surface, or volume depending on whether the space is one, two, or three dimensional. The integral range is a reference element to assess the magnitude of the **field** relative to the spatial continuity. In practice, a necessary condition to achieve **ergodicity** is to have a field that is large compared to the integral range. Whenever the integral range is infinite, there is no ergodicity.

INTERIOR. An interior point of a set of points S is a point P such that there is a **neighborhood** of P consisting entirely of points of S.

INTERMESHED STRUCTURE. An alternative term for **nested structure**.

INTERPOLATION. The process of finding a value of a **function** $f(x)$ at x_0, based on information provided by surrounding known values, $f(x_i)$. The **estimation** procedure must be something other than the law given by the function $f(x)$ itself. The term *extrapolation* is used instead when the location x_0 is outside the convex hull of the **data** locations, x_i, especially if there is a possibility that the function may change in form from the zone of x_i to the zone of x_0.

INTERQUARTILE RANGE. An alternative term for **quartile deviation**.

INTRINSIC. In **geostatistics**, the quality of honoring the **intrinsic hypothesis**.

INTRINSIC ATTRIBUTE. In a **probabilistic** model, an intrinsic attribute is any property that depends only on the **intrinsic random function** and not on the **modeling** of the intrinsic random function under study. For instance, the **generalized covariance** and an **estimator** from **kriging** are intrinsic attributes, but the **drift** is not.

INTRINSIC CORRELATION. Two or more **random functions** show intrinsic correlation if all the **covariances** and **cross-covariances** are proportional. Any of these random functions or its linear combinations have **autokrigeability**.

INTRINSIC DISPERSION. The slope of the **de Wijs semivariogram**.

INTRINSIC FUNCTION. An alternative term for **semivariogram**.

INTRINSIC HYPOTHESIS. **Weak stationarity** of the increments of a **random function**. These increments can be regarded as **permissible combinations of order 0**. It is a less restrictive form of **stationarity** than the comparable stationarity of order 2, and allows for the **modeling** of **regionalized variables** whose spatial **variance** apparently increases with **domain** size.

INTRINSIC KRIGING. A procedure for **estimating** a **regionalized variable** that is not **homogeneous** and is modeled by a **random function** with **permissible combinations of order** v that are **weakly stationary**. This form of the intrinsic kriging system of equations is identical to that of **universal kriging**, using a **polynomial of degree** v to **model** the **drift**. The only difference is in the inference of the required **generalized covariance**. In universal kriging, the **covariance** is inferred directly from the original **data**. The simplest approach calls for **structural analysis** in subareas of the entire **field**, analysis only along the strike, or both. Inference of the generalized covariance is done by determining the parameters of parametric covariance models through **cross-validation**.

INTRINSIC RANDOM FUNCTION OF ORDER v. A **random function** whose **generalized increments of order** v, $\tilde{Z}(\alpha)$, have **weak stationarity** and 0 **mean**. The **domain** of α includes the **permissible coefficients of order** v that have the same **distribution function** as the **translation of permissible coefficients** by any **vector**.

INVERSE OF A MATRIX. The inverse for a nonsingular **matrix** A is the **matrix** $\boldsymbol{A}^{-1}$ which satisfies

$$\boldsymbol{A}^{-1}\boldsymbol{A} = \boldsymbol{A}\boldsymbol{A}^{-1} = \boldsymbol{I}$$

where $\boldsymbol{I}$ is the **identity matrix**.

INVERSE PROBLEM. Determination of the parameters in a **model** in order to generate a response that duplicates the available observations. In general, the solutions to inverse problems are not unique.

ISOFACTORIAL BIVARIATE DENSITY FUNCTION. An alternative term for **diagonal bivariate density function**.

ISOLINE. A synonym for **isopleth**.

ISOPLETH. In a graphic display of a **function**, a line connecting points of equal value. Lines on a meteorological chart that connect locations experiencing the same temperature are examples of isopleths.

ISOTOPIC. A **sample** is isotopic if all **regionalized variables** in a **coregionalization** are known at all sample locations. Isotopic sampling significantly simplifies the study of coregionalizations, particularly **factorial kriging** analysis.

ISOTROPY. (a) The characteristic of a phenomenon in a **euclidean space** $\Re^n$ of dimension larger than 1 to have similar features in any dimension of $\Re^n$. (b) In **geostatistics**, the characteristic of a **structural function** to depend only on the modulus $|\boldsymbol{h}|$ of the separation **vector** $\boldsymbol{h}$, and not on its direction.

JACKKNIFE. (a) The jackknife, $J(\theta)$, is a statistical **estimator** of an unknown parameter θ of the form

$$J(\theta)=\frac{1}{m}\sum_{i=1}^{m}J_i(\theta)$$

where

$$J_i(\theta)=m\Theta-(m-1)\Theta_i$$

m is the number of subsets in a **partition** of the **data** set, Θ is an estimator of θ using the entire data set, and Θ_i is an estimator of the same parameter, but computed excluding the ith partition from the data set. For a **sample size** n, there are n/m estimators Θ_i, each one consisting of $(mn-n)/m$ observations. The jackknife method was introduced by Quenouille to reduce **bias**, arguing that if m is made equal to the sample size and Θ has a bias proportional to $1/n$, then the bias of $J(\theta)$ is proportional to $1/n^2$. Tukey extended the application of the jackknife by using Θ and $J(\theta)$ to approximate **confidence intervals** in certain **estimation** problems in which theoretical complexity or lack of reliable detailed knowledge about the **model** precludes a simpler theoretical approach. The jackknife procedure has been used to **estimate** confidence intervals for a **semi-variogram**; m should be sufficiently small so that each subset in the partition has at least 20 observations. (b) An improper synonym for **cross-validation**.

JOINT CROSS-VALIDATION. Cross-validation for two or more **regionalized variables** defined or **sampled** over the same **field**.

JOINTLY FACTORABLE RANDOM FUNCTION. A generalization of a **factorable random function** for the case of **coregionalizations**. Let $Y(\boldsymbol{x})$ and $Z(\boldsymbol{x})$ be two **random functions** defined over an n-dimensional **euclidean space** $\Re^n$. Let $g_i(\psi_i)$ and $g_j(\zeta_j)$ be **univariate probability density functions** for the **random variables** $Y(\boldsymbol{x}_i)$ and $Z(\boldsymbol{x}_j)$, respectively, and let $g(\psi_i,\zeta_j)$ be their corresponding **bivariate** density function. Let $\{p_k(\psi_i)\}$ and $\{q_k(\zeta_j)\}$ be two **complete sets of orthogonal polynomials** in $L_2[\Re^n,g(\psi_i)]$ and $L_2[\Re^n,g(\zeta_j)]$, respectively; namely, they are square **integrable functions with respect to $g_i(y_i)$ and $g_j(z_j)$**. The random functions $Y(\boldsymbol{x})$ and $Z(\boldsymbol{x})$ are called jointly factorable random functions if

$$\theta(\psi_i,\zeta_j)=\frac{g(\psi_i,\zeta_j)}{g(\psi_i)g(\zeta_j)}$$

is in $L_2[\Re^n\Re^n,g(\psi_i)g(\zeta_j)]$, and

$$h_m(\psi_i)=\int_{\Re}\theta(\psi_i,\zeta_j)g(\zeta_j)q_m(\zeta_j)d\zeta_j$$

$$h_m(\zeta_j)=\int_{\Re}\theta(\psi_i,\zeta_j)g(\psi_i)p_m(\psi_i)d\psi_i$$

are polynomials of a degree up to m. Jointly factorable random functions are a special case of **jointly orthogonal random functions**.

JOINTLY ORTHOGONAL RANDOM FUNCTION. (a) Two **random functions** $Y(\boldsymbol{x})$ and $Z(\boldsymbol{x})$ are called *jointly orthogonal* if

$$\mathrm{E}[Y(\boldsymbol{x}_i)Z(\boldsymbol{x}_j)]=0$$

whether the locations $\boldsymbol{x}_i$ and $\boldsymbol{x}_j$ are different or not. (b) A generalization of an **orthogonal random function** for the case of **coregionalizations**. Let k and l be nonnegative integers. Two random functions $Y(\boldsymbol{x})$ and $Z(\boldsymbol{x})$ are called *jointly orthogonal* if **complete sets of orthogonal pol-**

ynomials $\{p_k(\psi_i)\}$ and $\{q_l(\zeta_j)\}$ can be defined with respect to $g_i(\psi_i)$ and $g_j(\zeta_j)$ such that

$$\mathrm{E}[p_k(Y_i)q_l(Z_j)] = \delta_l^k\, \mathrm{E}[p_k(Y_i)q_k(Z_j)]$$

where δ_l^k is the **Kronecker delta.**

JOINTLY ORTHOGONAL RANDOM PROCESSES. Jointly orthogonal random functions along a line.

JOINT PROBABILITY DENSITY FUNCTION. A **probability density function** involving more than one **random variable.** It is equal to the partial derivative of nth order of the **joint probability distribution** $F(x_1,x_2,\ldots,x_i,\ldots,x_n)$ for **continuous random variables** when such a function exists.

$$f(x_1,x_2,\ldots,x_i,\ldots,x_n) = \frac{\partial^n F(x_1,x_2,\ldots,x_i,\ldots,x_n)}{\partial x_1,\partial x_2,\ldots,\partial x_i,\ldots,\partial x_n}$$

The **bivariate distribution** is a particular case of joint probability density function for two random variables.

JOINT PROBABILITY DISTRIBUTION. A **distribution function** involving two or more **random variables**

$$F(x_1,x_2,\ldots,x_i,\ldots,x_n) = P(X_1 \le x_1, X_2 \le x_2,\ldots,X_i \le x_i,\ldots,X_n \le x_n)$$

If there is a **joint probability density function** $f(x_1,x_2,\ldots,x_i,\ldots,x_n)$, it is related to the joint probability function by

$$f(x_1,x_2,\ldots,x_i,\ldots,x_n) = \frac{\partial^n F(x_1,x_2,\ldots,x_i,\ldots,x_n)}{\partial x_1,\partial x_2,\ldots,\partial x_i,\ldots,\partial x_n}$$

KRIGE'S ELLIPSE. A **scattergram** in which **estimates** of the **regionalized variable** appear along the abscissa and the values appear along the ordinate. The term ellipse refers to the characteristic elliptical scatter of points when the **bivariate distribution** of the true and estimated variables is **bigaussian**. The name honors D. G. Krige, who, during the prehistoric days of **geostatistics**, noted the relationship between estimated ore grades Z_v^* of blocks based on peripheral **samples** and observed grades Z_v of the blocks as these are mined out.

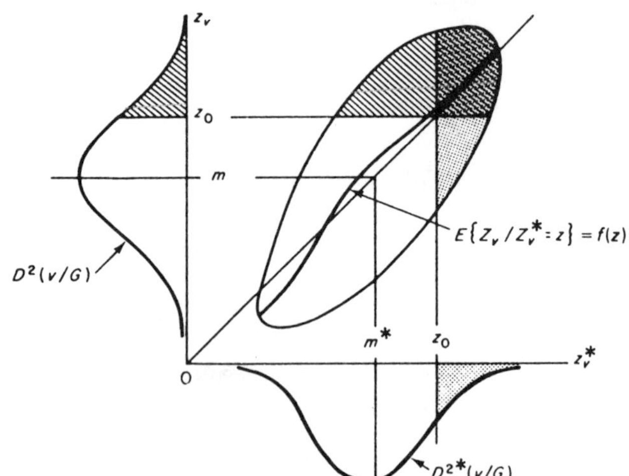

Copyright source: Journel and Huijbregts, 1978

KRIGE'S FORMULA. Let $v_1 \subset v_2 \subset v_3$ be three **supports** within the **field** of a **regionalized variable**. Krige's formula provides the **dispersion variance** $\mathrm{D}^2(\cdot)$ of the smallest volume v_1 in the largest volume v_3

$$\mathrm{D}^2(v_1|v_3) = \mathrm{D}^2(v_1|v_2) + \mathrm{D}^2(v_2|v_3)$$

Since all the terms in the formula are positive, $\mathrm{D}^2(v_1|v_3) \ge \mathrm{D}^2(v_2|v_3)$. In addition, since $v_1 \subset v_2$, unless there is a **hole effect**, the dispersion variance decreases as the support size increases.

KRIGING. A collection of generalized linear **regression** techniques for minimizing an **estimation variance** defined from a prior **model** for a **covariance** or a **generalized covariance of order** v that may or may not have constraints to eliminate **bias**. In contrast to classical linear regression, kriging takes into account **stochastic dependence** among the **data**. Among the most widely used forms of linear kriging are **simple kriging, ordinary kriging, universal kriging,** and **intrinsic kriging**. Nonlinear kriging is a linear regression on some appropriate nonlinear **transformation** of the original data and includes **lognormal kriging, multigaussian kriging, indicator kriging, probability kriging,** and **disjunctive kriging**. All kriging algorithms result in systems of linear equations involving **structural function** models and sometimes additional linear constraints. These systems of equations correspond to the **normal equations** of classical regression. Kriging algorithms can also be regarded as normal or orthogonal projections of the unknown **random variable** onto some **linear manifold** of the **Hilbert space** generated by the **data**, hence the name normal equations. Kriging is an **exact interpolator**. The spatial continuity of a kriged surface is greater than reality because of the **smoothing** effect inherent in the form of the **estimator**, which in most applications may be its worst drawback. The **estimation** technique is named after D.G. Krige, in honor of his empirical studies of South African gold mines that contributed to the formulation of the technique.

KRIGING NEIGHBORHOOD. That portion of a **field** in which the search for observations necessary to perform **kriging** can take place.

KRIGING VARIANCE. The kriging variance of a **random function**, $Z(\boldsymbol{x})$, is the nonconditional minimized **estimation variance** $\mathrm{E}[(Z(\boldsymbol{x}_0)-Z^*(\boldsymbol{x}_0))^2]$ resulting from a linear **kriging** process, or the **norm** of the **error** if kriging is expressed in terms of projections. The kriging **variance** is **homoscedastic**, as it is independent of the **data** used to obtain the **estimator**, $Z^*(\boldsymbol{x}_0)$. The kriging variance is a **measure** of **accuracy** of the **local** kriging **estimate** $z^*(\boldsymbol{x}_0)$ only in the special case when the random function **model** is a **multinormal distribution**. In all other cases, the kriging variance is only a ranking index of the data **configuration** showing which arrangement is, on **average**, better than others, without regard to individual **sample** values.

KRONECKER DELTA. The **function** δ^i_j of two variables i and j such that the function is 1, if $i = j$, and 0, if $i \neq j$. The Kronecker delta is useful for working with **orthogonal functions**.

KURTOSIS. A **measure** of peakedness or concentration around the **mode**. The degree of kurtosis is **measured** by the **coefficient of kurtosis** and the **coefficient of excess**.

LAG. (a) In its original sense, the difference in the time of occurrence of two events in a **time series**. (b) By analogy, the distance $\boldsymbol{h}$ between the locations of two **random variables** in a **random field**. The term should be distinguished from the sampling interval a, which is the difference in time or location between successive observations in a regular sampling taken along a traverse. A lag that is equal to the sampling interval is a special case representing a lag that is a multiple, na, of the sampling interval. Lags are not restricted to multiples of the sampling interval, but in general can be any distance in $\Re^n$.

LAGRANGE'S METHOD OF MULTIPLIERS. A procedure for the optimization of a **function** of several variables $f(x_1,x_2,\ldots,x_i\ldots,x_n)$ subject to constraints $\phi_1=0,\phi_2=0,\ldots,\phi_j=0,\ldots,\phi_m=0$. The solution is found by equating to 0 the n partial derivatives of the unconstrained auxiliary function

$$f+\sum_{j=1}^{m}\mu_j\phi_j$$

with respect to x_i, regarding the *Lagrange multipliers* μ_j as constants. The resulting n equations plus the original m constraints provide a system of equations from which the unknown x_i and μ_j can be computed. For example, to find the maximum area of a rectangle whose perimeter is a constant k, it is necessary to find the maximum value of the product xy, subject to the constraint $2x+2y-k=0$. The auxiliary function in this case is $u=xy+\mu(2x+2y-k)$, resulting in

$$x+2\mu=0$$
$$y+2\mu=0$$
$$2x+2y=k$$

the system of equations whose solution is $\mu=-0.125k, x=y=0.25k$. Lagrange's method is used to solve the **estimation variance** minimization problem behind most forms of **kriging**.

LAGRANGE MULTIPLIER. See **Lagrange's methods of multipliers**.

LAGUERRE POLYNOMIALS. The class of **functions** $L_n(x)$ obtained by finding **orthogonal functions** over the interval $(0,\infty)$ with respect to the weight function $w(x)=e^{-x}$. For any n, Laguerre polynomials satisfy the **recursion** formula

$$(1+2n-x)L_n(x)-n^2L_{n-1}(x)-L_{n+1}(x)=0$$

The corresponding **Rodrigues formula** is

$$L_n(x)=e^x\frac{d^n}{dx^n}(x^ne^{-x})$$

The first few Laguerre polynomials are

$$L_0(x)=1$$
$$L_1(x)=-x+1$$
$$L_2(x)=x^2-4x+2$$

It can be proven that

$$\sum_{n-1}^{\infty}L_n(x)\frac{t^n}{n!}=(1-t)^{-1}e^{\frac{-xt}{1-t}}$$

These polynomials are sometimes used in **disjunctive kriging** for the approximation of **distribution functions**.

LAPLACE TRANSFORM. The Laplace transform of a **function** $f(x)$ is the function $F(t)$ defined by

$$F(t)=\int_0^{\infty}e^{-tx}f(x)dx$$

where x is a real number, but t may be complex.

LAW OF LARGE NUMBERS. There are various expressions of the laws of large numbers, but the essential idea is the same for all. If the size of a **sample** is increased indefinitely or becomes very large, good sample **estimates** of the **population** parameters will tend to concentrate more and more closely about their true values. The law of large numbers states those conditions under which specific **functions** of **random variables** converge to constants as some parameter n—usually the **sample size**—tends to infinity. Strong expressions of the law are concerned with showing that, for example, a variable Θ converges to a value θ with unit **probability**. Weak expressions of the law consider conditions under which the probability that $|\Theta - \theta| \geq \varepsilon$ tends to 0. **Ergodicity** may be regarded as a form of the law of large numbers applied to **stationary** processes.

LEAST SQUARES METHOD. A **regression** of y on several variables $x_1, x_2, \ldots, x_i, \ldots, x_m$ has the form

$$y = f(x_1, x_2, \ldots, x_i, \ldots, x_m)$$

A **sample** may consist of a series of observations $\{y_1, y_2, \ldots, y_j, \ldots, y_n\}$ for the corresponding $x_{1j}, x_{2j}, \ldots, x_{ij}, \ldots, x_{mj}$. If the number of observations n exceeds the number of parameters k in the **function** $f(\cdot)$, then no **curve fitting** honoring all the observations is generally possible. The method of least squares provides an approximate fit by minimizing the sum of squares of the **residuals** $[y_j - f(x_{1j}, x_{2j}, \ldots, x_{ij}, \ldots, x_{mj})]$ over all n observations. The method yields reasonable approximations in many applications, and has optimal properties in some circumstances. For example, in the linear regression equation of the form

$$y = \beta_0 + \beta_1 x_1 + \ldots + \beta_i x_i + \ldots + \beta_m x_m + \varepsilon$$

where ε is a random residual having a **normal distribution** without **autocorrelation**, the **estimators** of β_i derived by least squares are **unbiased**, have minimum **estimation variance**, and are equivalent to **maximum likelihood estimators**. The simplest case is the **regression line**.

LEPTOKURTIC. See **coefficient of kurtosis**.

LIKELIHOOD FUNCTION. A **function** quantifying the **probability** of obtaining the **sample** $\boldsymbol{X}$ given parameters θ_i. Let $\{X_1, X_2, \ldots, X_j, \ldots, X_m\}$ be a random sample from a **population** whose **distribution** is known to be of a certain general form, but certain parameters are unknown. Let $f(\boldsymbol{X}; \theta_1, \theta_2, \ldots, \theta_i, \ldots, \theta_n)$ be the **frequency function** or the **probability density function** when the parameters have the values $\theta_1, \theta_2, \ldots, \theta_i, \ldots, \theta_n$, with θ_i restricted to the set Ω. The likelihood function $L(\cdot)$ is defined by

$$L(X_1, X_2, \ldots, X_j, \ldots, X_m; \theta_1, \theta_2, \ldots, \theta_i, \ldots, \theta_n) = \prod_{j=1}^{m} f(X_j; \theta_1, \theta_2, \ldots, \theta_i, \ldots, \theta_n)$$

and is considered to be a function of $\{\theta_1, \theta_2, \ldots, \theta_i, \ldots, \theta_n\}$ with $\{X_1, X_2, \ldots, X_j, \ldots, X_m\}$ fixed.

LINEAR COMBINATION OF VARIABLES. Given n variables $X_1, X_2, \ldots, X_i, \ldots, X_n$, their linear combination is the sum

$$\sum_{i=1}^{n} a_i X_i$$

in which a_i are constants, not all of them equal to 0.

LINEAR DEPENDENCE. Let $p_1, p_2, \ldots p_i, \ldots, p_n$ be a set of objects, such as **vectors**, **matrices**, or **functions**. The set of objects is said to have a linear dependence on another set $\alpha_1, \alpha_2, \ldots, \alpha_j, \ldots, \alpha_m$ if there is a combination such that

$$\sum_{i=1}^{n} \alpha_i p_i = 0$$

and not all α_i are 0. For example, let $f^i(\boldsymbol{x})$ be a set of functions with i an integer larger than 0 and less than the positive integer n. Let $\boldsymbol{x}_j$ be a set of points in a **euclidean space**. The set of functions $f^i(\boldsymbol{x})$ is said to have linear dependence for the set of points $\boldsymbol{x}_j$ with respect to the set of coefficients c_i, if for all j

$$\sum_{i=1}^{n} c_i f^i(\boldsymbol{x}_j) = 0$$

and at least one of the coefficients c_i is not equal to 0.

LINEAR FILTERING. An alternative term for **convolution**.

LINEAR GEOSTATISTICS. That part of **geostatistics** devoted to the study of linear **functions** of **regionalized variables**. When used in a **probabilistic** approach, linear geostatistics requires only the first two **moments** of the associated **random function**.

LINEAR INDEPENDENCE. A quality or state of not having **linear dependence**.

LINEAR MANIFOLD. A subset of a **vector space** restricted by one or more conditions on the weights of linear combinations of its members. **Ordinary kriging** can be regarded as the projection of an unknown **random variable** on a linear manifold.

LINEAR MODEL OF COREGIONALIZATION. Let $Z_i(\boldsymbol{x})$ be a collection of **random functions** in a **coregionalization**. There is a linear model of coregionalization if all the random functions in the coregionalization can be written as

$$Z_i(\boldsymbol{x}) = \sum_{l} a_i^l \, Y_l(\boldsymbol{x})$$

where a_i^l are real numbers and $Y_l(\boldsymbol{x})$ are **stochastically independent** random functions. This model is convenient for use in **factorial kriging**. Unfortunately, not every coregionalization can be represented as a linear model of coregionalization.

LINEAR SEMIVARIOGRAM. The non**transitive** type of **model**, $\gamma(h) = Ch$. The **Brownian motion process** has a linear semivariogram.

LINEAR SPACE. A set of **vectors** for which there is defined a product of multiplication of a vector by a real or complex number and an addition of any pair of vectors in the set. The operations of addition and multiplication must satisfy a certain number of axioms.

LOCAL. (a) Of, or relating to, a small portion of the **field** of a **regionalized variable**. (b) Of, or relating to, a **working neighborhood**.

LOCAL ESTIMATION. An **estimation** procedure that uses only a limited portion of the **field** of a **regionalized variable**. Local **estimates** are based only on a fraction of the available **data**.

LOCAL RELATIVE SEMIVARIOGRAM. A **linear combination** of **relative semivariograms** calculated over **local** portions of the **field**. If $\overset{*}{\gamma}_i(\boldsymbol{h})$ represents the **experimental semivariogram** for the ith **partition** out of a total of k partitions, $\overset{*}{m}_i$ is an **estimate** for the **mean** within the same partition, and $n_i(\boldsymbol{h})$ is the number of pairs used in the calculation of $\overset{*}{\gamma}_i(\boldsymbol{h})$, then an estimate for the local relative semivariogram is

$$\overset{*}{\gamma}_{LR}(\boldsymbol{h}) = \frac{1}{\sum_{i=1}^{k} n_i(\boldsymbol{h})} \sum_{i=1}^{k} \frac{n_i(\boldsymbol{h}) \, \overset{*}{\gamma}_i(\boldsymbol{h})}{\overset{*}{m}_i^{2}}$$

The partition of the field may pose a problem when the **sample size** is small, an inconvenience not shared by the **general relative semivariogram.**

LOGARITHMIC KRIGING. Kriging of a logarithmic **transformation** of the **data**. If the **random function** is **lognormally distributed**, the **estimation** has **bias** because the antilogarithm of a **transform** is an **estimate** of the **median** and not of the **mean**. If this bias is undesirable, then an alternative is to use **lognormal kriging.**

LOGARITHMIC SEMIVARIOGRAM. (a) The non**transitive** model

$$\gamma(h) = C\log_e(h)$$

Note that the logarithmic **model** cannot be used for distances smaller than $h=1$. The *de Wijsian semivariogram* is a special case of the logarithmic semivariogram that arises from a slight change in notation by making $C=3\alpha$. The parameter α is called the *absolute dispersion*. This model has the interesting property that, if a **support**, v_1, is similar in shape and enclosed in v_2, the **dispersion variance** $D^2(v_1|v_2)$ is given directly by

$$D^2(v_1|v_2) = \alpha\log_e(\frac{v_2}{v_1})$$

The computational advantage of this direct expression of the dispersion variance has disappeared with the development of computer programs that allow rapid calculation of the dispersion variance for any type of **semivariogram model**. Moreover, since $\gamma(h)<0$ for $h<1$ and $\gamma(0)=-\infty$, the logarithmic semivariogram can be used only after **regularization** over a nonzero volume. For these reasons, this model has fallen into disuse. (b) The **semivariogram** of the logarithmic **transformation** of the **data** $z(\cdot)$

$$\gamma(h) = \mathrm{E}[(\log_e z(\boldsymbol{x}) - \log_e z(\boldsymbol{x}+\boldsymbol{h}))^2]$$

This **structural function** is required to solve the **lognormal kriging** system of equations. The model fitted to the **experimental semivariogram** that results from the use of logarithmically transformed data can be any semivariogram model that satisfies the **permissibility criteria** and is not limited to the logarithmic model of definition (a).

LOGNORMAL CORRECTION. A **support effect** correction procedure that **transforms** the values of an original **lognormal distribution** to those of another **distribution** having the same **mean** m but a **variance** f times smaller. If $q_s(p)$ is a **quantile** for the distribution on the smaller **support** and $q_l(p)$ is the corresponding quantile on a larger support, then the transformation is given by

$$q_l(p) = a\,q_s^b(p)$$

where

$$a = \frac{m}{\sqrt{1+fC^2}}\left[\frac{\sqrt{1+C^2}}{m}\right]^2$$

$$b = \sqrt{\frac{\log(1+fC^2)}{\log(1+C^2)}}$$

f is the **variance adjustment factor** and C is the **coefficient of variation**. The lognormal transformation does not necessarily preserve the mean if it is applied to a distribution that is not lognormal. The drawback is solved by the **indirect lognormal correction.**

LOGNORMAL DISTRIBUTION. A **random variable** X is lognormally distributed if its natural logarithmic **transform** Y is **normally distributed** with **mean** μ_Y and **variance** σ_Y^2. The **probability density function** for X is a continuous **distribution** with a positive **coefficient of skewness** and is given by

$$f(x)=\begin{cases}0 & , \text{if } x \le 0\\ \dfrac{1}{\sqrt{2\pi}\sigma_Y x}e^{-\frac{[\log_e(x)-\mu_Y]^2}{2\sigma_Y^2}} & , \text{if } x>0\end{cases}$$

with

$$\mu_X = e^{\mu_Y + 0.5\sigma_Y^2}$$

$$\sigma_X^2 = \mu_X^2 (e^{\sigma_Y^2} - 1)$$

thus

$$\mu_Y = \log_e(\mu_X) - 0.5\,\sigma_Y^2$$

$$\sigma_Y^2 = \log_e\left(1 + \frac{\sigma_X^2}{\mu_X^2}\right)$$

LOGNORMAL KRIGING. **Simple kriging** or **ordinary kriging** of a logarithmically **transformed** set of observations having a **multivariate lognormal distribution**. Let $Y^*(\boldsymbol{x}_0)$ be the simple kriging **estimator** of the transformed **regionalized variable**, and let $\sigma_{E,Y}^2(\boldsymbol{x}_0)$ be the corresponding **estimation variance**. The **minimum mean square error estimator** $Z^*(\boldsymbol{x}_0)$ of the original variable and its corresponding estimation variance $\sigma_{E,Z}^2(\boldsymbol{x}_0)$ are

$$Z^*(\boldsymbol{x}_0) = e^{Y^*(\boldsymbol{x}_0)+0.5\sigma_{E,Y}^2(\boldsymbol{x}_0)}$$

$$\sigma_{E,Z}^2(\boldsymbol{x}_0) = m_Z^2\, e^{\sigma_Y^2}\left(1 - e^{-\sigma_{E,Y}^2(\boldsymbol{x}_0)}\right)$$

where m_Z^2 is the **mean** of the original regionalized variable, and σ_Y^2 is the **variance** for the logarithmic transform. The method does not have the **bias** of **logarithmic kriging**, but is very sensitive to small changes in the **logarithmic semivariogram** type of **model** required to compute $Y^*(\boldsymbol{x}_0)$ and $\sigma_{E,Y}^2(\boldsymbol{x}_0)$.

LOSS FUNCTION. A **function** that measures the impact of any particular level of **error** in the profitability of an operation. The **expected value** of a loss function has been proposed as a generalization of the current practice of using the **mean square error** as the **objective function** in the solution of **estimation** problems.

LOWER QUARTILE. The 0.25 **quantile**.

LOWER-UPPER (LU) DECOMPOSITION. An algorithm originally developed to find the **inverse of a matrix** and now used in **geostatistics** for the **conditional simulation** of **random fields**. Consider the set $S = S_1 \cup S_2$, where S_1 is the **sample** set and S_2 is the set of grid nodes used to define the simulated surface. Conveniently ordering the locations in S, the **covariance matrix** corresponding to S can be decomposed into four submatrices: C_{11}, the covariance matrix between the observations; $C_{12} = C_{21}$, the covariance matrices between the observations and the grid points; and C_{22}, the covariance matrix between grid points. Since the covariance matrix corresponding to S is a **symmetric matrix** that is positive-definite, it can be decomposed into the product of lower-upper triangular matrices

$$\begin{bmatrix} C_{11} & C_{12} \\ C_{12}^T & C_{22} \end{bmatrix} = \begin{bmatrix} L_{11} & 0 \\ U_{12}^T & L_{22} \end{bmatrix} \begin{bmatrix} U_{11} & U_{22} \\ 0 & U_{22} \end{bmatrix}$$

Let Z_1 be the **vector** of **data** values at the sample locations, and N_2 a vector of as many independent random numbers distributed according to a **standard normal distribution** as there are elements in S_2. A conditional simulation Y_2 at the locations in S_2 of the **random field** sampled by Z_1 is given by

$$Y_2 = U_{12}^T L_{11}^{-1} Z_1 + L_{22} N_2$$

Y_2 reproduces the covariance matrix associated with Z_1. To generate several **realizations** of the same random field, all that is required is to change the vector N_2. Applications currently are restricted to limited **sample sizes**. The use of moving data search **neighborhoods** is recommended for applications to large data sets.

MADOGRAM. A **measure** of the degree of dissimilarity in a **regionalized variable** as the distance between the observations being compared increases. The madogram is a **variogram of order 1**

$$\mathrm{E}[|Z(\boldsymbol{x}) - Z(\boldsymbol{x} + \boldsymbol{h})|]$$

The madogram is a continuous **function** of the **lag** rather than a single numerical value. Madogram **estimators** are more **resistant** to **outliers** than are **experimental semivariograms** and less sensitive to small **sample sizes**. The madogram can be used instead of the **semivariogram** in the **kriging** system of equations, resulting in a **norm** for the **error** rather than the traditional **estimation variance**. Like the estimation variance, the minimum error norm is a rank index of the **configuration** of the **data**.

MAIN DIAGONAL OF A MATRIX. The diagonal from the upper left corner to the lower right corner of a **square matrix**.

MANIFOLD. (a) In general, any collection or set of objects. (b) Smoothly curved sets of **functions** of dimension $m \leq n$ in an n-dimensional **euclidean space**. In statistics, manifolds are used to represent **sampling spaces** and parameter spaces for families of **probability functions**.

MARGINAL DISTRIBUTION. The **distribution function** of a **marginal probability density function**.

MARGINAL PROBABILITY DENSITY FUNCTION. Let $f(x_1,x_2)$ be the **joint probability density function** of two **random variables** X_1 and X_2. For the continuous case, the marginal probability density function for X_1 is the integral

$$f_1(x_1) = \int_{-\infty}^{\infty} f(x_1,x_2)\, dx_2$$

and for the discrete case

$$f_1(x_1) = \sum_{x_2} f(x_1,x_2)$$

There is a similar marginal probability density function for X_2. These two marginal probability density functions are uniquely defined by the joint probability density function. The converse, however, is not true. Given the marginal probability density functions for X_1 and X_2, it is generally possible to find several different corresponding joint probability density functions. Marginal probability density functions can be generalized to joint probability density functions and have an important role in statistical inference based on **Bayes' theorem**.

MARKOV PROCESS. A **stochastic process** in which a **prediction** is determined solely by the closest n observations, and is **stochastically independent** from all remaining more distant observations.

MATRIX. A rectangular tabulation of elements used to facilitate the study of problems in which the relationships between these elements are important, as in the solution of simultaneous linear equations. Matrices are commonly denoted by bold capital Roman letters; the individual elements are shown arranged inside square brackets

$$\mathbf{A} = \begin{bmatrix} a_{11} & a_{12} & a_{13} \\ a_{21} & a_{22} & a_{23} \end{bmatrix}$$

Unlike **determinants**, individual matrices do not have any quantitative value, but algebraic operations defined on matrices may lead to a single value or to another matrix.

MAXIMUM ENTROPY FORMALISM. See **entropy**.

MAXIMUM LIKELIHOOD ESTIMATOR. An alternative term for **maximum likelihood statistic.**

MAXIMUM LIKELIHOOD STATISTIC. Let $L(x_1, x_2, \ldots, x_n; \theta)$ be a **likelihood function** for a random **sample** of size n, and let $u(x_1, x_2, \ldots, x_n)$ be a nontrivial **function** of the sample. If $u(x_1, x_2, \ldots, x_n)$ is such that

$$L[x_1, x_2, \ldots, x_n; u(x_1, x_2, \ldots, x_n)] \geq L(x_1, x_2, \ldots, x_n; \theta) \quad \text{for every } \theta \in \Omega$$

then the **statistic** $u(X_1, X_2, \ldots, X_n)$ is called a maximum likelihood statistic. For instance, let $X_1, X_2, \ldots, X_n$ be a random sample from a **normal distribution** with **variance** 1 and **mean** μ, $-\infty < \mu < \infty$. In this particular case,

$$L(x_1, x_2, \ldots, x_n; \mu) = \left(\frac{1}{\sqrt{2\pi}}\right)^n e^{-0.5\sum_{i=1}^{n}(x_i - \mu)^2}$$

Following standard practice in calculus of applying **Lagrange's method of multipliers**, this function can be maximized by setting the first derivative of L with respect to μ equal to 0 and solving the resulting equation for μ under the constraint that the second derivative must be negative. However, since the maxima for L and $\log_e(L)$ occur for the same value of μ, it is easier in this case to find the maximum for the latter, which is

$$\mu^* = u(X_1, X_2, \ldots, X_n) = \frac{1}{n}\sum_{i=1}^{n} X_i$$

This is the unique maximum likelihood statistic for the mean μ. In this instance the statistic is **unbiased**, but in general, maximum likelihood statistics can have **bias**.

MEAN. The **expected value** $\mathrm{E}[X]$ of a **random variable** X. The mean represents the weighted **average** of a random variable when the weights are equal to the **probabilities** of occurrence. The mean is a **measure** of the central tendency of a random variable.

MEAN SEMIVARIOGRAM. Let v_x and v_y be two **supports**, at least one of them not a point. If x is a point inside v_x and y is a point inside v_y, then the **mean** value $\bar{\gamma}(v_x,v_y)$ for all possible point **semivariograms** $\gamma(x-y)$ is given by

$$\bar{\gamma}(v_x,v_y)=\frac{1}{v_x v_y}\int_{v_x} dx \int_{v_y} \gamma(x-y)dy$$

In practice, $\gamma(\cdot)$ can be obtained from tables or by numerical integration. The mean semivariogram is a **function** of both distance and shape of the volumes v_x and v_y.

MEAN SQUARE ERROR. The **expected value** of $(\theta^*-\theta)^2$, where θ^* is the **estimator** of the parameter θ.

MEASURE. (a) In ordinary use, a standard of comparison. (b) In connection to a set X and a **σ-algebra** $\mathfrak{m}$ of subsets of X, a measure $\mu(\cdot)$ is a **function** that assigns a certain nonnegative number to each set in $\mathfrak{m}$ and satisfies the following conditions:

i. If ϕ is the empty set, then

$$\mu(\phi)=0$$

ii. If A and B are two members of $\mathfrak{m}$ such that $A\cap B=\phi$, then

$$\mu(A\cup B)=\mu(A)+\mu(B)$$

iii. For any member A_i of $\mathfrak{m}$

$$\mu\left(\bigcup_{i=1}^{\infty}A_i\right)\le\sum_{i=1}^{\infty}\mu(A_i)$$

Definition (a) of **indicator** is an example of measure of a set.

MEDIAN. The 0.5 **quantile** M_d that divides evenly the **distribution function**, that is

$$P(X<M_d)=P(X\ge M_d)=\tfrac{1}{2}$$

MEDIAN INDICATOR KRIGING. An alternative term for **single indicator kriging**.

MEIJERING'S CELL. A synonym for **Voronoi polygon**.

MESOKURTIC. See **coefficient of kurtosis**.

METRIC SPACE. A set such that, for each pair of its elements (x,y), there is a **measure** for the degree of closeness between x and y. This measure is a nonnegative real **function** $d(x,y)$, called the distance between x and y, which satisfies

$$d(x,y)=0,\text{ if and only if } x=y$$
$$d(x,y)=d(y,x)$$
$$d(x,z)\le d(x,y)+d(y,z)$$

A **euclidean space** is a metric space.

MICROERGODICITY. A criterion used to determine which **model** parameters can be determined from knowledge of a single **realization** of a **random function** over a finite spatial **domain**, or a **regionalized variable**. Let $z(x)$ be the regionalized variable defined over a set S of an n-dimensional **euclidean space** and let $Z(x)$ be an associated **intrinsic random function** defined over the same set. Let $\theta[z(x)]$ be a **local** parameter, and let $\Theta[Z(x)]$ be the **random variable** that results from replacing the regionalized variable by the random function in the expression of the parameter. The parameter $\theta[z(\boldsymbol{x})]$is said to have microergodicity if

$$\frac{\mathrm{Var}[\Theta(Z(\boldsymbol{x}))-\theta(z(\boldsymbol{x}))]}{\theta[z(\boldsymbol{x})]^2} \leq \varepsilon$$

where ε is a small number. If $\theta[z(\boldsymbol{x})]$ possesses microergodicity, then its value is determined from only a single realization $z(\boldsymbol{x})$ of the random function. Otherwise, $\theta[z(\boldsymbol{x})]$ is a **conventional parameter**. Under certain weak hypotheses, microergodicity is preserved even in the non**stationary** case. Microergodicity is completely different from **ergodicity**. For example, in the case of a stationary and ergodic model, neither the **mean** nor the **variance** have microergodicity. However, in most instances, a **semivariogram** near its origin has microergodicity, the **de Wijsian semivariogram** being one of the exceptions.

MINIMUM ERROR NORM ESTIMATOR. An **estimator** that is best in the sense that it minimizes some predefined **norm** for the **error**, such as the **estimation variance**.

MINIMUM MEAN SQUARE ERROR ESTIMATOR. An **estimator** that is best in the sense that it minimizes the **mean square error**.

MODE. A local maximum of the **probability density function** of a **random variable**. See also **bimodal distribution, multimodal distribution**, and **unimodal distribution**.

MODEL. A representation of some or all of the properties of a device, system, or object for the purpose of deducing effects or better understanding the underlying process responsible for certain observations. A model can be conceptual, physical, or mathematical. Disagreement between observations and responses derived from a model demonstrates that the model imperfectly represents the actual situation, but unfortunately the converse is not true. Different combinations of parameters in a model can be mutually compensating and yield identical results, thus preventing a unique causal relationship.

MOMENT. A **moment of a distribution** about 0.

MOMENT OF A DISTRIBUTION. The **expected value** of a power, if it exists,

$$\mathrm{E}[(X-a)^k]$$

where k is an integer larger than 0. If a is the **mean**, then the moment is called the central moment of order k. Moments about 0 are simply called **moments**. If $x_1, x_2, \ldots, x_i, \ldots, x_n$ are values in a **sample** of n observations, then the central moments of the sample are the **statistics**

$$m_k = \frac{1}{n}\sum_{i=1}^{n}(x_i - m)^k$$

where m is the **arithmetic mean** of the observations. The first moment is the mean and the second moment about the mean is the **variance**.

MONOSCOPIC. Of, relating to, or involving a single objective. In opposition to **panscopic** scientific **theories**, most approaches used in **geostatistics** are monoscopic, in the sense that what is valued in the **model** is its efficiency in solving specific problems. No importance is attached to the behavior of the models in other situations that are not of direct interest.

MONTE CARLO METHOD. Any of a number of procedures that use **simulated** random numbers to obtain a **probabilistic** approximation to the solution of a mathematical or physical problem. With the exception that the methods do not require **data**, they are similar to traditional statistical procedures in which random **samples** are used to make inferences about actual **populations**. In many applications of the Monte Carlo method, the problem being studied does not have an explicit random component; then, a deterministic parameter is expressed as a parameter of some **distribution function** and that distribution is simulated. Use of the Monte Carlo method may result in a significant mathematical simplification of the original problem.

MOSAIC MODEL. A particular kind of **random function** in which all **indicator cross-covariances** possess **intrinsic correlation**. Under this **model**, the approximation of **single indicator kriging** is justified.

MOVING LINEAR AVERAGE. Given a sequence of **random variables** $\{X_1, X_2, \ldots, X_i, \ldots, X_m\}$, the moving average of order k is the sequence $\{Y_1, Y_2, \ldots, Y_i, \ldots, Y_{m-k+1}\}$, where

$$Y_i = \frac{1}{k} \sum_{j=0}^{k-1} w_{i+j} X_{i+j}$$

and the w_i represent weights.

MULTIGAUSSIAN KRIGING. The process of determining the **conditional distribution** of a **random function** whose **normal score transform** is assumed to follow a **multinormal distribution**. The random function is thus fully characterized by the **covariance** of these normal score transforms. The process involves **simple kriging** of the normal score data, and a restoration of the conditional distribution of the original random function by a **Monte Carlo method**. Multigaussian kriging is a generalization of **lognormal kriging** in the sense that it does not require any hypothesis about the **marginal probability density function** of the original variables.

MULTIMODAL DISTRIBUTION. A **distribution** with more than two **modes**. Such a distribution often results from mixing together observations from several **populations** into a single **sample**.

MULTINORMAL DISTRIBUTION. A random **vector** $\boldsymbol{X}$ consisting of two or more **random variables** X_i is said to have a multinormal distribution if and only if every linear combination of its n components X_i has a **univariate** normal distribution. The **joint probability distribution** of these components is fully characterized by the **mean** vector

$$\boldsymbol{m} = [\mathrm{E}(X_1), \mathrm{E}(X_2), \ldots, \mathrm{E}(X_i), \ldots, \mathrm{E}(X_n)]$$

and the n by n **covariance matrix**

$$\boldsymbol{K} = \mathrm{E}[(\boldsymbol{X} - \boldsymbol{m})(\boldsymbol{X}^T - \boldsymbol{m}^T)]$$

where the superscript T denotes the **transpose of the matrix** which is the base of T. The **conditional distribution** given any number of **realizations** of the remainder components is a **normal distribution** whose conditional mean is equal to the **estimate** of X_i provided by **simple kriging**, and a **homoscedastic** conditional **variance** equal to the corresponding simple **kriging variance**. **Stationary** random variables that have a multinormal distribution are fully characterized by their **covariances** and constant mean.

MULTIPLE INDICATOR KRIGING. **Simple kriging** or **ordinary kriging** applied to **indicator data columns**. Multiple indicator kriging uses a different **model** of the indicator **variogram** for each threshold considered.

MULTIVARIATE. Involving two or more **random variables**.

MULTIVARIATE DISTRIBUTION. An alternative term for **joint probability distribution**.

NEGATIVE BINOMIAL DISTRIBUTION. Let X be a **discrete random variable** that can assume any nonnegative integer value, let p be a number larger than 0 but less than 1, and let k be a real or integer positive number. The random variable X has a negative binomial distribution if its **probability function** is given by

$$P(X=x)=\frac{\Gamma(x+k)}{\Gamma(k)x!}p^k(1-p)^x$$

where $\Gamma(\cdot)$ is the **gamma function**. The **mean** is $k(1-p)/p$ and the **variance** $k(1-p)/p^2$. In a **Bernoulli experiment**, the negative binomial distribution can be used to **model** the number of failures x of an **outcome** whose **probability** of occurrence is p, before the kth success.

NEIGHBORHOOD. The ε-neighborhood of a given point P is the **open** region comprising the set of all points whose distance from P is less than ε, such as the **interior** of a circle in the plane with the point as center.

NESTED SEMIVARIOGRAM. A linear combination of two or more **models** of **semivariograms** that satisfy the **permissibility criteria**, such as the sum of a **pure nugget effect semivariogram model** plus two **spherical semivariograms** with different parameters. A pure nugget effect model plus another model is not always regarded as a nested model. Instead, it is referred to as a model with a **nugget effect,** such as an **exponential semivariogram** with a nugget effect.

NESTED STRUCTURE. Let $Z(\boldsymbol{x})$ be a **random function** modeling a **regionalized variable** whose spatial continuity is the compound effect of several genetic sources of spatial variation. A nested structure is each one of the additive components in the random function **model** $Z(\boldsymbol{x})$ that results in the observed **experimental semivariogram**. The larger the number of nested structures, the more complex the model of the spatial variability of the regionalized variable. Nested structures can be obtained by a process of **curve fitting** to the experimental semivariogram by trial and error. The idea of nested structures is used in **factorial kriging** to split the random function into several components.

NONCENTERED COVARIANCE. Let X and Y be two **random variables** having a **joint probability density function** and **means** μ_X and μ_Y, respectively. The noncentered covariance is the **moment**

$$\mathrm{E}[XY]$$

The noncentered and the **centered covariance** are related through

$$\mathrm{E}[(X-\mu_X)(Y-\mu_Y)]=\mathrm{E}[XY]-\mu_X\mu_Y$$

The noncentered covariance can be regarded as an **inner product**.

NONERGODIC EXPERIMENTAL COVARIANCE. The **covariance** of a **random field** $Z(\boldsymbol{x})$ defined by

$$\mathrm{Cov}^*(\boldsymbol{h})=\frac{1}{n(\boldsymbol{h})}\sum_{i=1}^{n(\boldsymbol{h})}[z(\boldsymbol{x}_i)-m_1^*(\boldsymbol{h})][z(\boldsymbol{x}_i+\boldsymbol{h})-m_2^*(\boldsymbol{h})]$$

$$m_1^*(\boldsymbol{h})=\frac{1}{n(\boldsymbol{h})}\sum_{i=1}^{n(\boldsymbol{h})}z(\boldsymbol{x}_i)$$

$$m_2^*(\boldsymbol{h})=\frac{1}{n(\boldsymbol{h})}\sum_{i=1}^{n(\boldsymbol{h})}z(\boldsymbol{x}_i+\boldsymbol{h})$$

where $n(\boldsymbol{h})$ is the number of pairs a distance $\boldsymbol{h}$ apart. This nontraditional covariance **estimate** is non**ergodic** in the sense that it does not implicitly assume that

$$m_1^*(\boldsymbol{h}) \approx m_2^*(\boldsymbol{h}) \approx m = \mathrm{E}[Z(\boldsymbol{x})], \text{ for all } \boldsymbol{h}$$

NONLINEAR GEOSTATISTICS. That part of **geostatistics** devoted to the study of nonlinear **functions** of a given **regionalized variable**. A simple example of a nonlinear function is an **indicator**. Nonlinear geostatistics is used especially to solve mining selection problems. As the nature of the problems becomes more complex, nonlinear geostatistics also provides a simpler approach to solve linear problems.

NONPARAMETRIC GEOSTATISTICS. A branch of **geostatistics** comprising techniques that, although they are of a **probabilistic** nature, do not rely on a prior **multivariate distribution** hypothesis about the **random function** used to **model** the **regionalized variables**. Instead, nonparametric geostatistics draws from the **data** as much as possible before calling for a model. **Indicator kriging** and **probability kriging** are examples of such techniques.

NONSINGULAR SYSTEM. An alternative term for **regular system**.

NORM. In working with **functions**, **vectors**, or **matrices**, the form of the problem may suggest a way to ascribe a nonnegative number to each of the objects as a **measure** of its magnitude, in a way similar to the modulus of a complex number. Such a numerical measure associated with the elements x and y of a **vector space** is denoted by $\|x\|$ and is called a norm, provided that the following properties are satisfied:

$$\text{(a) } \|x\| \begin{cases} = 0, & \text{if } x = 0 \\ > 0, & \text{for } x \neq 0 \end{cases}$$

$$\text{(b) } \|\lambda x\| = |\lambda| \, \|x\|$$

$$\text{(c) } \|x + y\| \leq \|x\| + \|y\|$$

where λ is any real or complex number. Norms are important in studying convergency properties of sequences of vectors and matrices, and for obtaining **error** bounds. The square norm of an error vector does not need to be its **variance**.

NORMAL DISTRIBUTION. The **variate** X is a normal **random variable** when the **probability density function** is given by

$$g(x) = \frac{1}{\sigma\sqrt{2\pi}} e^{-\frac{(x-\mu)^2}{2\sigma^2}}, \quad -\infty < x < \infty$$

where μ is the **mean** and σ^2 the **variance**. The probability density function is bell-shaped, and is completely determined by the mean and the variance. First formulated by de Moivre in 1733, and later studied and named after several others, the only other surviving synonym is *Gaussian distribution*.

NORMAL EQUATIONS. Any system of equations arising from the application of the **least squares method**, and more generally, from the **projection theorem**. Since the solution to the equation set can be regarded as a process of normal or orthogonal projection onto some **linear manifold** of the **Hilbert space** generated by the **data**, the equations are referred to as normal equations. In **geostatistics**, such equations are usually the system of n linear equations that are the solutions to the generalized **regression estimate** z^* of a value based on n observations z_i that may have **stochas-**

tic dependence

$$z^* - \mathrm{E}[z] = \sum_{i=1}^{n} \lambda_i (z_i - \mathrm{E}[z_i])$$

The solution leads to the same system of equations as **simple kriging** and yields the weights λ_i. If some of the **expected values** are not known, linear constraints on the weights are added to ensure that the **estimator** is **unbiased**. **Ordinary kriging** and **universal kriging** are two examples of constrained systems of normal equations.

NORMAL PROBLEM. A synonym for **direct problem**.

NORMAL SCORE TRANSFORM. See **Gaussian anamorphosis**.

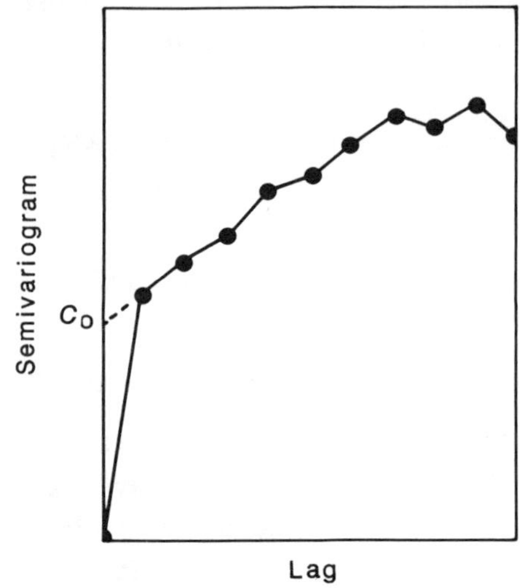

NUGGET EFFECT. An apparent discontinuity in the **experimental semivariogram** near the origin caused by measurement **errors** or by **nested structures** that have **ranges** smaller than the sampling interval, or both. Theoretically, the value of the **semivariogram** must be 0 at the origin. However, in the presence of a nugget effect, the semivariogram does not seem to approach 0 at the origin, but rather some positive value C_0 that is significantly larger than 0. **Kriging** using **semivariogram models** having a nugget effect leads to apparent discontinuities in the **predictions**. Although kriging still honors the **data** when the estimation location $\boldsymbol{x}_0$ coincides exactly with the location $\boldsymbol{x}_i$ of an observation, if $\boldsymbol{x}_0 \neq \boldsymbol{x}_i$, then $z^*(\boldsymbol{x}_0)$ is significantly different from $z(\boldsymbol{x}_i)$, no matter how close the locations.

OBJECTIVE FUNCTION. A **measure** of effectiveness in the optimization of systems. The **estimation variance** is the objective function for **kriging**.

OBJECTIVE MEANING. The property of being able to distinguish from among the various concepts, parameters, and statements in a **model** those that have in the reality of the physical world a counterpart, which is observable and measurable. Parameters without an objective meaning are called **conventional parameters**.

OBJECTIVITY CRITERION. Any of the necessary and sufficient conditions for a **model** parameter that will allow its **experimental** verification. In **geostatistics**, this criterion is specified by the *regional characteristic principle*, which states that objective parameters in a model are only those that can be described in terms of **regional variables**. The principle is used to insure that computations have a physical meaning, thus justifying the use of **topoprobabilistic models** to characterize **regionalized variables**.

OPEN. A property of a set of points that denotes that all points in the set are **interior** points. An open set is the complement of a **closed** set. The interior of a circle and the set of all points on one side of a straight line in a plane are open sets.

OPERATIONAL CONCEPT. A notion defined by a regular, repeatable process that allows the **experimenter** to quantify or **measure** the process.

OPERATOR. (a) In its widest sense, operator is synonymous with **function**. (b) In a more restricted sense, the term operator denotes a function for which both the **domain** and the **range** consist of **vectors** or functions. For example, a differential operator produces the derivative of a function.

OPTIMAL RESIDUAL. A **residual** obtained by subtracting an optimal **estimate** of the **drift**, in some specific sense, from the original **random function**.

ORDER OF A DETERMINANT. The number of elements in either a **row** or a **column of a determinant.**

ORDER OF A SQUARE MATRIX. The number of elements in either a **column** or a **row** of a **square matrix.**

ORDINARY KRIGING. A linear **estimation** algorithm for a **regionalized variable** that satisfies the **intrinsic hypothesis.** In contrast to **simple kriging**, the **mean** is unknown. The form of the **estimator** $Z^*(\boldsymbol{x}_0)$ of the regionalized variable at location $\boldsymbol{x}_0$, using n observations taken at sites $\boldsymbol{x}_i$, is

$$Z^*(\boldsymbol{x}_0)=\sum_{i=1}^{n}\lambda_i\,Z(\boldsymbol{x}_i)$$

The λ_i are unknown weights that can be obtained by solving the system of equations

$$\begin{bmatrix}\lambda_1\\ \cdots\\ \lambda_i\\ \cdots\\ \lambda_n\\ \mu_0\end{bmatrix}=\begin{bmatrix}0 & \cdots & \gamma(\boldsymbol{x}_1,\boldsymbol{x}_i) & \cdots & \gamma(\boldsymbol{x}_1,\boldsymbol{x}_n) & 1\\ \cdot & \cdot & \cdot & \cdot & \cdot & \cdot\\ \gamma(\boldsymbol{x}_i,\boldsymbol{x}_1) & \cdots & 0 & \cdots & \gamma(\boldsymbol{x}_i,\boldsymbol{x}_n) & 1\\ \cdot & \cdot & \cdot & \cdot & \cdot & \cdot\\ \gamma(\boldsymbol{x}_n,\boldsymbol{x}_1) & \cdots & \gamma(\boldsymbol{x}_n,\boldsymbol{x}_i) & \cdots & 0 & 1\\ 1 & \cdots & 1 & \cdots & 1 & 0\end{bmatrix}^{-1}\begin{bmatrix}\gamma(\boldsymbol{x}_1,\boldsymbol{x}_0)\\ \cdots\\ \gamma(\boldsymbol{x}_i,\boldsymbol{x}_0)\\ \cdots\\ \gamma(\boldsymbol{x}_n,\boldsymbol{x}_0)\\ 1\end{bmatrix}$$

where μ_0 is a **Lagrange multiplier** arising from the optimization method used to insure that the estimator is **unbiased** and has minimum **mean square error.** $\gamma(\cdot)$ is the **semivariogram model** for the **random function** $Z(\boldsymbol{x})$. The **kriging variance,** $\sigma_E^2(\boldsymbol{x}_0)$, is given by

$$\sigma_E^2(\boldsymbol{x}_0)=\mu_0+\sum_{i=1}^{n}\lambda_i\,\gamma(\boldsymbol{x}_i,\boldsymbol{x}_0)$$

The method is related to **interpolation** by **spline functions.**

ORTHOGONAL FUNCTION. Any of the **functions** $\{f_k(x)\}$ in a set of functions is said to be orthogonal with respect to a nonnegative function $g(x)$ over the interval (a,b) if

$$\int_a^b f_i(x)\,f_j(x)\,g(x)\,dx=\alpha\,\delta_j^i$$

where i and j are any nonnegative integers, δ_j^i is a **Kronecker delta,** and α is a constant. Orthogonal functions provide a powerful alternative to the direct approach of approximating a **data** set by the **least squares method.**

ORTHOGONAL POLYNOMIAL. Any of the **orthogonal functions** for the particular case when the set of **functions** $\{f_k(x)\}$ is a set of polynomials $\{p_k(x)\}$.

ORTHOGONAL RANDOM FUNCTION. (a) A **random function** $Z(\boldsymbol{x})$ for which

$$\mathrm{E}[Z(\boldsymbol{x}_i)Z(\boldsymbol{x}_j)]=0$$

for any two different locations $\boldsymbol{x}_i$ and $\boldsymbol{x}_j$. (b) Let k and l be nonnegative integers. A random function $Z(\boldsymbol{x})$ is called orthogonal if there exists a **complete set of orthogonal polynomials** $\{p_k(\psi_i)\}$ with respect to $g_i(\zeta_i)$ and $\{q_l(\zeta_j)\}$ with respect to $g_j(\zeta_j)$ such that

$$\mathrm{E}[p_k(Z_i)q_l(Z_j)] = \delta_l^k \, \mathrm{E}[p_k(Z_i)q_k(Z_j)]$$

where δ_l^k is the **Kronecker delta**. Orthogonal random functions form a broad class which includes **factorable random functions** as a special case.

ORTHOGONAL RANDOM PROCESS. An **orthogonal random function** on the line.

ORTHOGONAL VECTOR. Any of a pair of **vectors** whose **inner product** is 0. In two- or three-dimensional **euclidean spaces**, this is equivalent to the vectors being perpendicular.

ORTHONORMAL FUNCTION. Two **functions** $f_i(x)$ and $f_j(x)$ are said to be orthonormal if they are **orthogonal functions** and the constant α is equal to 1.

ORTHONORMAL VECTOR. Any of a pair of **orthogonal vectors** of unit length.

OUTCOME. The end product of a **trial** in a **random experiment**.

OUTLIER. In a **sample**, any of the few observations that are separated so far in value from the remaining measurements that the questions arise whether they belong to a different **population**, or that the sampling technique is faulty. The determination that an observation is an outlier may be highly subjective, as there are no strict criteria for deciding what is and what is not an outlier.

OVERESTIMATION. See **error**.

PAIRWISE RELATIVE SEMIVARIOGRAM. An **experimental semivariogram** in which every difference is adjusted by the **average value** of the pair of observations. For a **random field** $Z(\cdot)$ that is **intrinsic** or second order **stationary**, the pairwise semivariogram $\overset{*}{\gamma}_{PR}(\boldsymbol{h})$ is the **estimate**

$$\overset{*}{\gamma}_{PR}(\boldsymbol{h}) = \frac{1}{2n(\boldsymbol{h})} \sum_{i=1}^{n(\boldsymbol{h})} \left[\frac{z(\boldsymbol{x}_i + \boldsymbol{h}) - z(\boldsymbol{x}_i)}{0.5\{z(\boldsymbol{x}_i + \boldsymbol{h}) + z(\boldsymbol{x}_i)\}} \right]^2$$

where $n(\boldsymbol{h})$ is the number of pairs of observations a distance $\boldsymbol{h}$ apart. The pairwise relative semivariogram is less sensitive to **outliers** than to the experimental semivariogram.

PANSCOPIC. Of, relating to, or involving all possible objectives related to an endeavor. In general, given the number, size, and generality of the objectives pursued, most scientific **models** and **theories** are panscopic. Most methods used in **geostatistics** are not panscopic, but rather they are **monoscopic**.

PARABOLIC BEHAVIOR. A **function** is said to possess parabolic behavior if the function can be approximated by a **polynomial of degree 2**. **Structural functions** of highly continuous **regionalized variables**, such as those having a **trend**, have parabolic behavior near the origin.

PARABOLIC SEMIVARIOGRAM. The non**transitive** and non**positive definite** function

$$\gamma(\boldsymbol{h}) = C h^2$$

Regionalized variables having a linear **drift** produce parabolic **experimental semivariograms**. Since the **function** is not positive definite, it cannot be used as a **semivariogram model**.

PARSEVAL'S EQUATION. See **Bessel inequality**.

PARTITION OF A SET. A collection of sets v_i is said to be a partition of a set V if

i. $v_i \cap v_j = \varnothing$, for $i \neq j$

ii. $\cup v_{i=V}$

PEARSON CORRELATION COEFFICIENT. A synonym for **correlation coefficient.**

PENTA-SPHERICAL SEMIVARIOGRAM. The **transitive** type **model**

$$\gamma(h)=\begin{cases} C\left[\frac{15}{8}\left(\frac{h}{a}\right)-\frac{5}{4}\left(\frac{h}{a}\right)^3+\frac{3}{8}\left(\frac{h}{a}\right)^5\right] & , 0 \leq h < a \\ C & , a \leq h \end{cases}$$

where C is the **sill** and a is the **range**. The penta-spherical semivariogram is intermediate in form between the **spherical semivariogram** and the **exponential semivariogram.**

PERCENTILE. See **quantile.**

PERMISSIBILITY CRITERIA. The set of necessary or sufficient conditions—or both—that must be satisfied by a **model** in order to be used for a given purpose. In **geostatistics,** the permissibility criteria include the constraints on a **function** in order for it to be used as a **covariance, semivariogram,** or **generalized covariance.**

PERMISSIBLE COEFFICIENTS OF ORDER ν. A set of coefficients $\{\alpha_1, \alpha_2, \ldots, \alpha_i, \ldots, \alpha_n\}$, each one linked to a location $\boldsymbol{x}_i$, that **filters** out **polynomials of degree** $k \leq \nu$, $p_k(\cdot)$, that is

$$\sum_{i=1}^{n} \alpha_i \, p_k(\boldsymbol{x}_i) = 0, \quad \text{for all } k \leq \nu, \text{ and all } x_i$$

PERMISSIBLE COMBINATIONS OF ORDER ν. A linear combination of **random variables** with **permissible coefficients of order ν.** Given an **intrinsic random function or order ν,** permissible combinations of order ν are the only combinations that have **weak stationarity.**

PIVOT. Any of the **elements of a matrix** along the **main diagonal.**

PLATYKURTIC. See **coefficient of kurtosis.**

POINT KRIGING. Kriging in which both the observations and the **estimates** have the same **support,** as in the case when the supports are points. The **structural function** used in point kriging must be defined for the same support. Point kriging is a special kind of **interpolation** procedure.

POISSON DISTRIBUTION. A discrete random variable X is said to have a Poisson distribution if its probability function is given by

$$P(X=x)=\begin{cases} \frac{\lambda^x e^{-\lambda}}{x!} & , x=0,1,2,\ldots \\ 0 & , \text{elsewhere} \end{cases}$$

The **mean** and the **variance** of the Poisson distribution are both equal to λ. In a **Poisson process,** the number of occurrences of **events** in an interval of length t follows a Poisson distribution with mean νt, where ν is a positive constant characteristic of each process.

POISSON PROCESS. An integer-valued **stochastic process** that satisfies the following assumptions:

(a) $\{X(t), t \geq 0\}$ has **stationary** unitary independent increments.

(b) For any times r and s such that $r<s$, the number of counts $X(s)-X(r)$ in the interval from r to s is distributed according to a **Poisson distribution** with **mean** $\nu(s-r)$. **Realizations** of a Poisson process plot in the form of an unbounded staircase. The process can be generalized to n-dimensional **euclidean spaces**, where the counts are taken in volumes in $\Re^n$. The geographical distribution of some plants or animals can be **modeled** by two-dimensional Poisson processes.

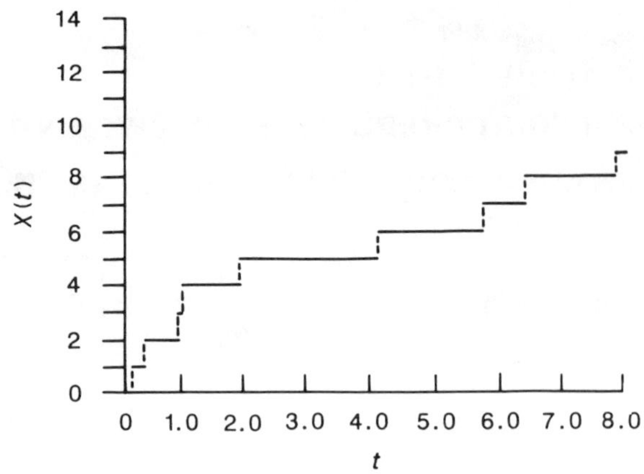

Copyright source: Parzen, 1965

POLYGON OF INFLUENCE. A synonymous term for **Voronoi polygon.**

POLYNOMIAL IN SEVERAL VARIABLES. An expression of the same structure as polynomials for one variable, except that the powers are products of more than one variable, such as $1.2-x-2xy^2$.

POLYNOMIAL OF DEGREE n. An integral, algebraic, and **rational expression** of one variable in the form

$$\sum_{i=0}^{n} a_i x^i$$

where a_i is a complex or real number, and n is a nonnegative integer, as in $3-x+0.5x^3$.

POLYSCOPIC. Of, relating to, or involving several objectives related to a given endeavor. A polyscopic approach is intermediate between **panscopic** and **monoscopic** undertakings.

POPULATION. (a) The set of all possible **outcomes** of a **random experiment**. (b) The set of all individuals of a predetermined group within which statistical inference is to be performed.

POSITIVE DEFINITE. (a) Let m be an integer larger than 0. Let $\{\alpha_1,\alpha_2,\ldots,\alpha_i,\ldots,\alpha_m\}$ be any set of real or complex numbers and let $\{x_1,x_2,\ldots,x_i,\ldots,x_m\}$ be any set of points in an n-dimensional **euclidean space** $\Re^n$. A continuous, symmetric **function** $f(\cdot)$ is said to be positive definite if:

$$\sum_{i=1}^{m}\sum_{j=1}^{m}\alpha_i\,\alpha_j\,f(x_i-x_j)\geq 0$$

A **covariance** function in an n-dimensional euclidean space must be positive definite in that space. The negative of a **semivariogram** in $\Re^n$ must be **conditionally positive definite of order 0**. That is, the preceding inequality must hold only for those weights that satisfy

$$\sum_{i=1}^{m}\alpha_i=0$$

Every linear combination of covariances or semivariograms with positive coefficients is also a covariance or semivariogram, and every function that is positive definite in an n-dimensional euclidean space is also positive definite in every space of a smaller dimension. The converse, however, is not true. The strictly positive definite nature of covariances and the negative of semivariograms is a necessary condition for the existence of a unique solution to the **kriging** system of equations and for nonnegative **estimation variances**. The condition is not sufficient because even though a function can be positive definite, the system of equations may still be singular for a given set of **data**. This situation most commonly arises when more than one observation is taken at the same location. (b) A **Hermitian matrix** is said to be positive definite if all its **eigenvalues** are positive or 0.

POSTERIOR DISTRIBUTION. A **probability density function, probability function,** or **distribution function** that summarizes information about a **random variable** after having obtained information from empirical **data.** The use of the designation posterior distribution is favored within the context of the statistical methodology of Bayesian inference. In **nonparametric geostatistics,** the term **conditional distribution** is used as a synonym.

POWER SEMIVARIOGRAM MODEL. The non**transitive** model

$$\gamma(h) = Ch^{\alpha}, \ 0 < \alpha < 2$$

The **linear semivariogram** is a special case of this **model** when $\alpha = 1$.

PRECISION. The degree of approximation with which a quantity is stated. In an **experiment,** precision expresses the consistency in a set of measurements, regardless of whether the observations conform to the true value or not. A precise measurement is not necessarily **accurate.** Generally, precision is proportional to the **standard deviation** of the **sample** and shows in the number of significant digits in a reported value. For instance, a device measuring a concentration as 9.1280 percent is more precise than another reporting 9.3 percent.

PREDICTION. (a) In **time series** analysis, the forecasting of a **realization** of a **stochastic process** at a point beyond the most recent observation. (b) In a **regression** equation that expresses the variability of a **random variable** Y in terms of other random variables $X_1, X_2, \ldots, X_n$, the expected value taken by Y for given values $x_1, x_2, \ldots, x_n$. In **geostatistics,** the term is used in this sense interchangeably with definition (b) for **estimation.**

PREPOSTERIOR DISTRIBUTION. A **distribution function** of **local** nature that summarizes supplementary information about a **random variable** at a particular location. This knowledge is available before further information is obtained from empirical **data,** and is in addition to the **prior distribution** of the **random function** used to **model** the **regionalized variable.** As opposed to the prior distribution, preposterior distributions are not **stationary** as they change from one location to another. Examples of supplementary information leading to preposterior distributions include knowledge of the degree of alteration or rock type in a certain area. Preposterior distributions are treated as **fuzzy indicator data** in the process of **multiple indicator kriging.**

PRINCIPAL COMPONENT. Given a **covariance matrix** of n **random variables** possessing **stochastic dependence,** a principal component is any of the **eigenvectors** X_{λ_i}, where the λ_i are the **eigenvalues.** The principal components appear as n **orthogonal vectors** with decreasing **variances** that are linear combinations of the original random variables. **Kriging** the principal components and then **transforming** the results back into **estimators** of the original random variables captures the **cross-correlation** information carried by the original covariance matrix.

PRINCIPAL DIAGONAL OF A MATRIX. An alternative term for **main diagonal of a matrix.**

PRIOR DISTRIBUTION. A **distribution function, probability density function,** or **probability function** that summarizes information about a **random variable** or parameter as known or assumed before obtaining further information from empirical **data.** The term is primarily used within the context of Bayesian inference. The notion of subjective formulation for prior distributions has been the matter of acrimonious debates.

PROBABILISTIC REGIONAL SEMIVARIOGRAM. The **random function** $\Gamma_R(\boldsymbol{h})$ that is obtained when a **regionalized variable** $z(\boldsymbol{x})$, defined over S, is replaced by its associated random function $Z(\boldsymbol{x})$ in the expression of the regional semivariogram $\gamma_R(\boldsymbol{h})$

$$\Gamma_R(\boldsymbol{h})=\frac{1}{2V(\boldsymbol{h})}\int_{S(\boldsymbol{h})}[Z(\boldsymbol{x}+\boldsymbol{h})-Z(\boldsymbol{x})]^2\,d\boldsymbol{x}$$

where $\boldsymbol{h}$ is any **vector** in S, the intersection of S with its translation by $\boldsymbol{h}$ is denoted $S(\boldsymbol{h})$, and $V(\boldsymbol{h})$ is the integral

$$V(\boldsymbol{h})=\int_{S(\boldsymbol{h})}d\boldsymbol{x}$$

Under the **intrinsic hypothesis**, the **expected value** of this probabilistic regional semivariogram is equal to the **theoretical semivariogram**.

PROBABILITY. A basic concept that has been expressed in different ways, all fortunately leading to much the same results. (a) The axiomatic definition. The probability of an **event** $\mathcal{A}_i$ is a number assigned to this event. This number satisfies the following three postulates, but is otherwise unspecified:

i. The probability $P(\mathcal{A}_i)$ is nonnegative

$$P(\mathcal{A}_i)\geq 0$$

ii. The probability of the **sample space** Ω is 1

$$P(\Omega)=1$$

iii. For any sequence of events $\{\mathcal{A}_1,\mathcal{A}_2,\ldots,\mathcal{A}_i,\ldots\}$ that are mutually exclusive

$$P\left(\bigcup_{i=1}^{\infty}\mathcal{A}_i\right)=\sum_{i=1}^{\infty}P(\mathcal{A}_i)$$

(b) The relative frequency definition. Let an **experiment** be repeated n times. If the event $\mathcal{A}$ occurs $n_{\mathcal{A}}$ times, then its probability $P(\mathcal{A})$ is defined as

$$P(\mathcal{A})=\lim_{n\to\infty}\frac{n_{\mathcal{A}}}{n}$$

Although definition (b) is popular among physicists and engineers because it provides some physical meaning to computed probabilities, it is strongly rejected by mathematicians who object to the idea of infinitely repeating a real experiment, and object to defining a limit by an experimental ratio. (c) Classical definition. Let $n_{\mathcal{A}}$ be the number of times that the event $\mathcal{A}$ can occur out of a total of n alternatives. Then $P(\mathcal{A})$ is equal to the ratio of the favorable alternatives to the total number of alternatives

$$P(\mathcal{A})=\frac{n_{\mathbf{A}}}{n}$$

According to this definition, the probability $P(\mathcal{A})$ is found a priori without actual experimentation. There are several shortcomings to this definition. Counting does not lead to the correct result unless an "equally likely" specification is added to the definition. The definition then becomes circular, as "equally likely" is used to mean "equally probable." In complex situations, the derivation of the ratio is difficult or impossible without the backing of some experimentation. Finally, in some circumstances the total number of **outcomes** is not finite. This definition does satisfy the axioms of definition (a). (d) The term probability also is used as a **measure** of belief that something may or may not happen. This is a subjective statement about a single event, closer to inductive reasoning, and not an **average value** about mass phenomena. Although widely used in applications of **Bayes' theorem**, this interpretation of probability is a radical departure from the concept used in the physical sciences.

PROBABILITY DENSITY FUNCTION. For a given a **probability function** $P(\cdot)$, a probability density function is a **function** $f(x)$ such that, for any **event** $\mathcal{A}$, the **probability function** $P(\mathcal{A})$ may be expressed as the integral

$$P(\mathcal{A}) = \int_{\mathcal{A}} f(x)dx$$

When the **domain** of $f(x)$ is a set of real numbers and $f(x)$ is continuous at x, then the probability density function is the derivative of the **distribution function**.

PROBABILITY DISTRIBUTION. A synonym for **probability function**.

PROBABILITY FUNCTION. Let Ω be a set and $\mathcal{F}$ be a **σ-algebra** of subsets A_i of Ω. A probability function is a set **function**, $P(\cdot)$, defined for every subset in $\mathcal{F}$ and which satisfies the following conditions

i. $P(A_i) \geq 0$, for each A_i in $\mathcal{F}$

ii. If $\{A_1, A_2, \ldots, A_i, \ldots\}$ is a sequence of pairwise disjoint subsets of $\mathcal{F}$, then

$$P\left(\bigcup_{i=1}^{\infty} A_i\right) = \sum_{i=1}^{\infty} P(A_i)$$

iii. $P(\Omega) = 1$

It follows from the definition that the probability function **range** is the **closed** interval $[0,1]$.

PROBABILITY INTERVAL. A nonparametric **measure** of dispersion in the values of a **random variable** given by the difference between any two conveniently specified **quantiles**. The **quartile deviation** is a particular case of probability interval.

PROBABILITY KRIGING. An enhancement of **indicator kriging** in which, in addition to **indicator** values for the same threshold, the **estimator** of **conditional probability** also includes **rank order transformations** of the original **data**. The objective of probability kriging is to provide **models** for the **conditional distribution** of the unknown attribute at any unsampled location.

PROBABILITY MEASURE. A synonym for **probability function**.

PROBABILITY SPACE. Let Ω be a **sample space** and let $\mathcal{F}$ be a **σ-algebra** of subsets E_i of Ω, for which a **probability function** $P(E_i)$ is defined. The probability space is the triplet $(\Omega, \mathcal{F}, P)$. Kolmogorov formulated the concept in the 1930s and used it first in the study of natural phenomena when he attempted **modeling** turbulence in fluid flow.

PRODUCT OF A SCALAR AND A MATRIX. Let λ be a constant and $\boldsymbol{A}$ a **matrix**. The product $\lambda\boldsymbol{A}$ of scalar λ and matrix $\boldsymbol{A}$ is a second matrix $\boldsymbol{B}$ whose elements are equal to the corresponding elements of $\boldsymbol{A}$ times λ.

PRODUCT OF MATRICES. Let $\boldsymbol{A}$ and $\boldsymbol{B}$ be two **matrices**, such that the number of columns in $\boldsymbol{A}$ is equal to the number of rows in $\boldsymbol{B}$

$$A = \begin{bmatrix} a_{11} & a_{12} & \dots & a_{1i} & \dots & a_{1n} \\ a_{21} & a_{22} & \dots & a_{2i} & \dots & a_{2n} \\ \cdot & \cdot & \dots & \cdot & \dots & \cdot \\ a_{l1} & a_{l2} & \dots & a_{li} & \dots & a_{ln} \\ \cdot & \cdot & \dots & \cdot & \dots & \cdot \\ a_{k1} & a_{k2} & \dots & a_{ki} & \dots & a_{kn} \end{bmatrix} \qquad B = \begin{bmatrix} b_{11} & b_{12} & \dots & b_{1j} & \dots & b_{1m} \\ b_{21} & b_{22} & \dots & b_{2j} & \dots & b_{2m} \\ \cdot & \cdot & \dots & \cdot & \dots & \cdot \\ b_{i1} & b_{i2} & \dots & b_{ij} & \dots & b_{im} \\ \cdot & \cdot & \dots & \cdot & \dots & \cdot \\ b_{n1} & b_{n2} & \dots & b_{nj} & \dots & b_{nm} \end{bmatrix}$$

The product ***A B*** of these two matrices is a third matrix whose elements are given by

$$c_{rs} = \sum_{i=1}^{n} a_{ri}\, b_{is}$$

Matrix ***C*** has the same number of rows as ***A*** and the same number of columns as ***B***.

PROJECTION THEOREM. Let A be a **Hilbert space**, let B be a **closed, linear space** included in A, and let x be an element of A. The projection theorem states that there exists one and only one element y in B that minimizes the distance to x. This unique element y is called the projection of x onto the subspace B.

PROPORTIONAL EFFECT. An **heteroscedastic** condition in which the **variance** of the **error** is proportional to some **function** of the **local** mean of the **data** values. **Random fields** having **log-normal distributions** always possess a proportional effect. Let X be a random field, let Y be its logarithmic **transformation**, and let σ_X^2 and σ_Y^2 be their respective variances. Since the variances are related by

$$\sigma_X^2 = \mu_X^2 (e^{\sigma_Y^2} - 1)$$

where μ_X is the **mean** of variable X if the **field** is subdivided, the variance σ_X^2 will be directly proportional to the square of the mean inside each of the subdivisions.

PROXIMAL POLYGON. A synonym for **Voronoi polygon**.

PUNCTUAL KRIGING. An alternative term for **point kriging**.

PURE NUGGET EFFECT. The property of possessing a **transitive** type of **semivariogram** having a **sill** equal to the **a priori variance** and a **range** of 0 or smaller than the sampling interval. **White noise** has pure nugget effect.

PURE NUGGET EFFECT SEMIVARIOGRAM MODEL. The **model**

$$\gamma(r) = C_0\, H(r)$$

where C_0 is a positive constant and $H(r)$ is the **Heaviside's function**.

PYTHAGOREAN THEOREM. See **Bessel inequality**.

Q-Q PLOT. A **scattergram** in which the **quantiles** of two series of observations are plotted. If the two series have the same **sample size**, then it is not necessary to calculate the quantiles, as the same results can be obtained by plotting the **rank order transforms** of the series.

QUANTILE. Any of the $(n-1)$ numbers that **partition a set** of values of a **random variable** so that the **cumulative frequency** of the **population** or **sample** is divided into n classes each containing the same number of observations. Most commonly used quantiles have special names. The quantile that divides a random variable into 2 classes is called the **median**; into 4 classes, they are *quartiles*; into 5 classes, they are quintiles; into 6 classes, they are sextiles; into 10 classes, they are *deciles*; and into 100 classes, they are *centiles* or *percentiles*.

QUARTILE. See **quantile**.

QUARTILE DEVIATION. A **measure** of **sample** dispersion based upon the distance between the **upper quartile** and the **lower quartile**.

QUASI-STATIONARY. Having **weak stationarity** within **neighborhoods** of the **local** type.

RANDOM COIN METHOD. An alternative term for **random tokens method**.

RANDOM EXPERIMENT. An **experiment** that does not always yield the same result when repeated under the same conditions.

RANDOM FIELD. An alternative term for **random function**.

RANDOM FUNCTION. Let D be a spatial **domain** in an n-dimensional **euclidean space** and let Ω be a **sample space**. A random function $\{Z(\boldsymbol{x},\omega):\boldsymbol{x}\in D,\ \omega\in\Omega\}$ is a collection of **random variables**. For a fixed value $\boldsymbol{x}_i$ of $\boldsymbol{x}$, $Z(\boldsymbol{x}_i,\omega)$ is a **function** of ω, that is, $Z(\boldsymbol{x}_i,\omega)$ is a random variable. In contrast, for a specific ω_j in Ω, $Z(\boldsymbol{x},\omega_j)$ is a deterministic function of $\boldsymbol{x}$ that represents a possible observation of the **random field**. The function $Z(\boldsymbol{x},\omega_j)$ is called a **realization** or **sample function** of the process. Finally, $Z(\boldsymbol{x}_i,\omega_j)$ is merely a number. To simplify the notation, the variable ω is generally omitted, $Z(\boldsymbol{x},\omega)$ is referred to simply as $Z(\boldsymbol{x})$, and $Z(\boldsymbol{x},\omega_j)$ as $z(\boldsymbol{x})$. **Geostatistics** relies heavily on random functions that have an imposed **spatial dependence** between the various component random variables for the **modeling** of **regionalized variables**.

RANDOM KRIGING. A variant of **simple kriging** or **ordinary kriging** where the location of the **data** is **uniformly distributed** within a given area or volume. The corresponding **semivariogram** distances are replaced by a **mean semivariogram** over the given space. However, the diagonal of the left-hand kriging **matrix** contains the actual location and **support** of the data. The resulting **estimates** are no longer the best, but the loss in **accuracy** is supposedly compensated for by the computational savings, especially for large data sets. This simplification is no longer widely used and was justified in preliminary surveys before the use of digital computers.

RANDOM PROCESS. A synonym for **stochastic process**. An alternative term for a **random function** in a one-dimensional **euclidean space**.

RANDOM SET. A **function** S assigning one set s to each element a of a **sample space**. In practice, interest is limited to **closed** sets. The rule is complemented by defining a *Choquet's capacity* $T(\cdot)$, which is the **probability** that a closed random set S has no intersection with an arbitrary **compact** set K.

$$T(K)=P(S\cap K=\varnothing)$$

The notion of random sets is a generalization of the idea of **random variables** and is used in the **simulation** of geometric shapes.

RANDOM TOKENS METHOD. A method used in the **simulation** of two-dimensional **Boolean random functions**. The procedure involves three steps. First, points are generated according to a **Poisson process** and these points are used as the centers of circles having radii of random lengths. Then, a **random function** is defined inside each circle; the function is constant for each circle and **stochastically independent** of variables inside other circles. Finally, for each point

$\boldsymbol{x}$ in $\Re^2$, the contributions of all circles containing $\boldsymbol{x}$ are added.

RANDOM VARIABLE. A **function** X assigning one and only one real number $X(a) = x$ to each element a of a **sample space**. It is understood that a **probability function** is also defined to account for the possibility that X takes certain values.

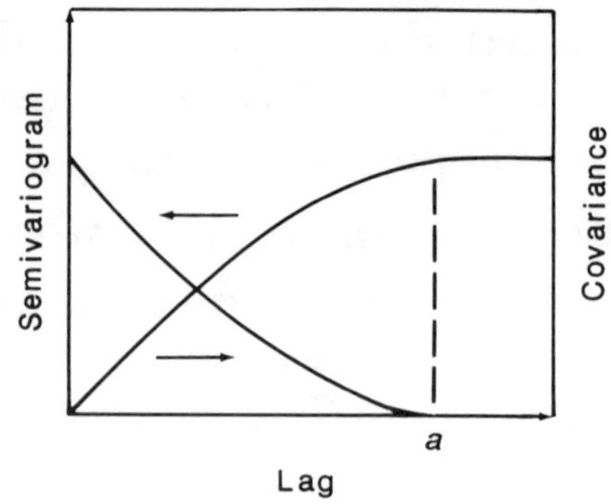

RANGE. (a) In the widest mathematical sense, the range is the set of possible values that a **function** or variable can assume. (b) In **geostatistics**, the maximum distance, a, separating pairs of **random variables** that **model** a **regionalized variable** that has any significant **statistical dependence**. For a **semivariogram** that is **transitive**, the range is the smallest semivariogram argument for which the semivariogram is either exactly equal to the **sill** or asymptotically close to the sill. In a transitive **covariance**, the range is the smallest abscissa for which the function is equal to 0, or asymptotically close to 0.

RANK ORDER TRANSFORM. Let z_i be a **sample** of n observations $\{z_1, z_2, \ldots, z_i, \ldots, z_n\}$. The sample can be ordered by increasing values, with r_i being the rank of observation z_i. For example, the rank of the observation having the smallest value is 1, and that having the largest value is n. The integer value **transform** $r_i = u(z_i)$ is called the rank order transform of the ith observation relative to the set $\{z_1, z_2, \ldots, z_i, \ldots, z_n\}$. Rank order transform **data** are used in **probability kriging** and to provide **structural functions** that are **resistant** to **outliers**.

RANK ORDER VARIOGRAM. Twice the **experimental semivariogram** of the **rank order transforms** of the original **data**. Because the rank order variogram is **resistant**, it is a convenient **measure** of spatial continuity for sparse data containing **outliers**. The rank order variogram allows detection of the **ranges** of any linear or nonlinear **spatial dependence** or patterns of **anisotropy**.

RATIONAL EXPRESSION. A **function** that does not include any variable in an irreducible radical or under a fractional exponent. A rational expression can be written as a quotient of polynomials.

REALIZATION. A set of values $\{z(\boldsymbol{x}_1), z(\boldsymbol{x}_2), \ldots, z(\boldsymbol{x}_i), \ldots, z(\boldsymbol{x}_n)\}$ that may arise from a **random function** $Z(\boldsymbol{x})$. This realization may be regarded as a member of the random function in the same way that an individual observation is regarded as a member of a **population**. In general, the realization is of infinite extent, but a **sample** of finite **sampling size** also is referred to as a realization.

RECOVERABLE RESERVES. That fraction of the ore **reserves in place** in a mine that is of economic worth. This notion involves both financial, technical, and economic considerations.

RECOVERY FUNCTION. Let V be a panel whose **partition** is made of smaller blocks v_i that will be used as selection units. A recovery function provides the amount of ore or the tonnage of metal relative to a cutoff grade. If selection is done on actual block grades, the recovery function is called the direct recovery function. If the selection is done on **estimated** block grades, the recovery function is called an indirect recovery function. The **estimation** of a direct recovery function is a **local estimation** problem without **information effect**. The estimation of an indirect recovery function is a problem in local estimation with **information effect**. This term has superseded the older **geostatistical** concept of **transfer function**.

RECURSION. The continuous repetition of the same operation or group of operations. A **function** is said to be recursive if it can be defined in terms of itself. For instance, for any positive integer n, given that $1!=1$, the calculation of $n!$ by recursion would be through $n!=n(n-1)!$.

REDUCED RESIDUAL. The ratio of the **error** of **kriging** over the positive square root of the corresponding **kriging variance** σ_E^2. If σ_E^2 **estimates** with **accuracy** the error **variance**, the variances of the reduced residuals of approximately similar **data** sets should be equal to 1. Reduced residuals have been used to automate **structural analysis**. If a **semivariogram model** $\gamma(\boldsymbol{h})$ is used in a **cross-validation** study and the variance of the resulting residual is $\overset{*2}{\sigma_E}$, then a second cross-validation study using the semivariogram model $\overset{*2}{\sigma_E}\,\gamma(\boldsymbol{h})$ will produce reduced residuals whose variance is exactly 1. However, such rescaling has no effect on the estimates and the procedure is suboptimal in the sense that it cannot account for alternative models.

REGIONAL CHARACTERISTIC. An attribute of a **regionalized variable** that could be calculated without **error** if the regionalized variable were **sampled** exhaustively in the entire **field**. Only regional characteristics have an **objective meaning**. An **average value** is a regional characteristic but an **expected value** is not.

REGIONAL CHARACTERISTIC PRINCIPLE. See **objectivity criterion**.

REGIONALIZED VARIABLE. A single-valued **function** defined over a **metric space**. In practice, a regionalized variable is preferentially used to describe natural phenomena, such as the elevation of the bottom of the sea, that are characterized by fluctuations that are smooth at a **global** scale but erratic enough at a **local** scale to preclude their analytical **modeling**. **Geostatistics** models regionalized variables as **realizations** of **random functions**.

REGIONAL MAGNITUDE. An alternative term for **regional characteristic**.

REGIONAL MAGNITUDE PRINCIPLE. A synonymous term for **regional characteristic principle**.

REGIONAL SEMIVARIOGRAM. Let $z(\boldsymbol{x})$ be a **regionalized variable** defined over a **field** S. The regional semivariogram is the **regional characteristic** defined by

$$\gamma_R(\boldsymbol{h})=\frac{1}{2V(\boldsymbol{h})}\int_{S(\boldsymbol{h})}[z(\boldsymbol{x}+\boldsymbol{h})-z(\boldsymbol{x})]^2\,d\boldsymbol{x}$$

where $\boldsymbol{h}$ is any **vector** in S, $S(\boldsymbol{h})$ is the intersection S with its translation by $\boldsymbol{h}$, and $V(\boldsymbol{h})$ is the integral

$$V(\boldsymbol{h})=\int_{S(\boldsymbol{h})}d\boldsymbol{x}$$

In a **topoprobabilistic model** this semivariogram can be regarded as a **realization** of the **probabilistic regional semivariogram**.

REGRESSION. Any of several statistical methods developed to investigate relationships between variables. A regression empirically provides the **conditional expectation** of a **random variable** by assuming an analytical expression for other variables and using **data** to determine the parameters, usually by a minimum **mean square error** procedure.

REGRESSION LINE. If the **conditional expectation** of a **random variable** Y, given the value of a random variable X, is given by

$$f(x)=a+bx$$

the graph of this equation is called the regression line. Provided the **least squares method** is used to compute the coefficients, it can be shown that the coefficients are given by

$$a=\mu_Y-\frac{\mathrm{Cov}(X,Y)}{\sigma_X^2}\mu_X$$

$$b=\frac{\mathrm{Cov}(X,Y)}{\sigma_X^2}$$

where μ_X and μ_Y are the **means** of X and Y, $\mathrm{Cov}(X,Y)$ is the **covariance** between the random variables, and σ_X^2 is the **variance** of X.

REGULARIZATION. In general, the act of making more uniform. Technically, the term is used with two different meanings. (a) Let $f(\boldsymbol{x})$ be a **function** to be **smoothed** and let $w(\boldsymbol{x})$ be a weighting function. The regularization of $f(\boldsymbol{x})$ by $w(\boldsymbol{x})$ is the **convolution** $f(\boldsymbol{x})*w(\boldsymbol{x})$. Regularization can be regarded as a form of **linear filtering** that has the general effect of **smoothing** small-scale fluctuations in $f(\boldsymbol{x})$. The weighting function is often taken as the **indicator** of a **support** centered at $\boldsymbol{x}$. The regularized function then is a linear **average value** of $f(\boldsymbol{x})$ over that support. (b) The replacement of a **sample** that is irregularly located in space by a regular **configuration**, either by resampling or **gridding**.

REGULARIZED SEMIVARIOGRAM. The **semivariogram** $\gamma_v(\boldsymbol{h})$ of a **regionalized variable** having a **support** that is not a point. A regularized semivariogram and a point semivariogram are related through

$$\gamma_v(\boldsymbol{h})=\bar{\gamma}(v,v_{\boldsymbol{h}})-\bar{\gamma}(v,v)$$

where $v_{\boldsymbol{h}}$ denotes the support translated by $\boldsymbol{h}$, and $\bar{\gamma}(\cdot)$ is the **mean semivariogram** for the point support. The figure is a graphical comparison between a point and a regularized semivariogram for the special case of a **spherical semivariogram**. The support for the regularized semivariogram is a cylindrical core of length l significantly longer than the cross-sectional radius.

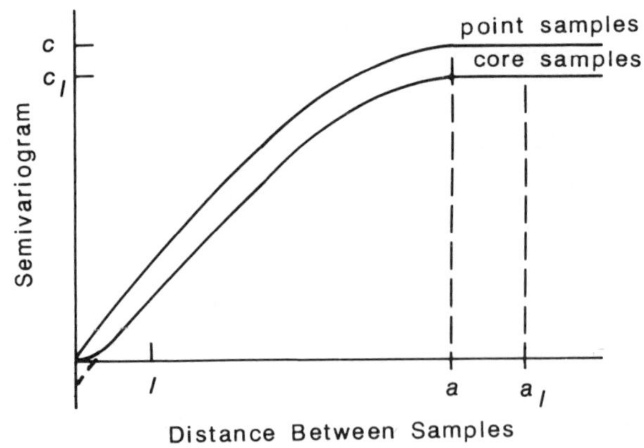

Copyright source: Clark, 1979

REGULAR SYSTEM. A system of linear equations having one and only one solution. For the **kriging** system of equations, the necessary and sufficient conditions for a regular system are: (1) The negative of the **semivariogram** or any **structural function** must be a **strictly conditionally positive definite function** for a given **data** set. (2) The **basis function** must have **linear independence** for such a data set.

RELATIVE NUGGET EFFECT. The ratio of the **nugget effect** constant to the **sill** in a **semivariogram model.**

RELATIVE SEMIVARIOGRAM. The relative semivariogram $\gamma_R(\boldsymbol{h})$ is a **semivariogram** $\gamma(\boldsymbol{h})$ scaled by some **function** $f(m^*)$ of either the **global** mean or some **estimate** of **local** mean values.

$$\gamma_R(\boldsymbol{h}) = \frac{\gamma(\boldsymbol{h})}{f(m^*)}$$

The relative semivariogram is sometimes used to cancel out the **proportional effect**. The most commonly used scaling function is the square of the **mean**, m^{*2}, which produces an adequate cancelation of the proportional effect only for **multinormal distributions**.

REPRESENTATION OF AN INTRINSIC RANDOM FUNCTION. Let $Z(\boldsymbol{x})$ be a non**stationary** type of **random function** defined on an n-dimensional **euclidean space**. A representation of an intrinsic random function $\tilde{Z}(\alpha)$ is any linear combination of components $Z(\boldsymbol{x}_i)$ weighted by **permissible coefficients of order v**

$$\tilde{Z}(\alpha) = \sum_{i=1}^{n} \alpha_i Z(\boldsymbol{x}_i)$$

Each $\tilde{Z}(\alpha)$ has many representations. The algebraic difference between two representations of the same intrinsic random function is a polynomial. In practice, a **regionalized variable** under study is regarded as a **realization** of a representation of an intrinsic random function.

RESERVES IN PLACE. The total quantity of a mineral resource actually present at its natural place of occurrence, regardless of economic or technical considerations related to its exploitation.

RESIDUAL. A general term denoting a quantity remaining after some other value has been subtracted. (a) The difference between a **random field** and its **drift**. (b) A synonym for **error** in an **experiment**.

RESIDUAL KRIGING. Simple kriging modified to account for a prior **estimate** of the **drift** $m(\boldsymbol{x})$ of a non**homogeneous** regionalized variable $z(\boldsymbol{x})$. A **global** estimate of the drift $m^*(\boldsymbol{x}_i)$ is first subtracted from each one of the observations $z(\boldsymbol{x}_i)$ to obtain **residuals** $r(\boldsymbol{x}_i)$. A final estimate $z^*(\boldsymbol{x}_0)$ is obtained by adding the corresponding drift estimate $m^*(\boldsymbol{x}_0)$ to a **kriging** estimate of the residual $r^*(\boldsymbol{x}_0)$. Global estimates of the drift may be obtained by a conventional **least squares method**; by a stepwise iterative generalized **regression** for simultaneously estimating the global drift and the **semivariogram** of **correlated** residuals; or directly from considerations based on the genesis of the **regionalized variable**. Only under very special circumstances can the estimates be identical to those obtained by **universal kriging**.

RESISTANT. The characteristic of a **statistic**, such as an **experimental semivariogram**, to be insensitive to the presence of anomalous **data** values, such as **outliers**. For example, the **rodogram** is more resistant to outliers than the traditional **semivariogram**.

ROBUST. The characteristic of a **model** to persist in yielding correct **predictions** in spite of **errors** in the collection of **data**, in the choice of a model, in the specification of the model, or any combination of these.

RODOGRAM. A **measure** of the degree of dissimilarity in the **outcome** of sampling as the distance between sites being compared increases. The rodogram is half of a **variogram of order 0.5**

$$0.5\,\mathrm{E}[|Z(\boldsymbol{x}) - Z(\boldsymbol{x}+\boldsymbol{h})|^{\frac{1}{2}}]$$

Since the increments are a **function** of the distance between the pair of locations being considered, the rodogram is a continuous function rather than a single numerical value. Rodograms are more **resistant** to **outliers** than are the more widely used **experimental semivariograms**, particularly when the **data** are sparse.

RODRIGUES FORMULA. A formula that expresses the nth **function** of a class of **orthogonal polynomials** $p_n(x)$ as the nth derivative of other functions

$$p_n(x)=\frac{1}{\rho(x)}\frac{d^n}{dx^n}\sigma(x)\rho(x)$$

where $\rho(x)$ is a polynomial and $\sigma(x)$ is an arbitrary function. For example, the Rodrigues formula for **Laguerre polynomials** is

$$L_n(x)=\frac{1}{e^{-x}}\frac{d^n}{dx^n}(x^n e^{-x})$$

ROSE DIAGRAM. A graphical display in polar coordinates. The graph of a discontinuous set of only a few values may consist of radiating rays whose lengths are proportional to the quantities being represented. Rose diagrams are used to detect and display **anisotropy** in **experimental semivariograms** and **experimental covariograms**.

ROW OF A DETERMINANT. A collection of elements along a horizontal line and in the same sequence as they appear in a **determinant**.

ROW OF A MATRIX. A collection of elements along a horizontal line and in the same sequence as they appear in a **matrix**.

σ-ALGEBRA. Let X be a set and let $\mathcal{F}$ be a nonempty collection of subsets that satisfies the following three axioms:

i. If ϕ is the empty set, ϕ belongs to $\mathcal{F}$.
ii. If a subset belongs to $\mathcal{F}$, its complement also belongs to $\mathcal{F}$.
iii. If $\{A_1, A_2, \ldots, A_i, \ldots\}$ is a sequence of members of $\mathcal{F}$, then their union $\bigcup_{i=1}^{\infty} A_i$ also belongs to $\mathcal{F}$.

These axioms are necessary to deal with an infinite number of **events**. The set of all subsets of X satisfying these three axioms is called the total σ-algebra. For a given set X there may be several different σ-algebras. A σ-algebra is a fundamental concept in the mathematical description of random events in a **sample space**.

SAMPLE. The term is used with two related but slightly different meanings. (a) In statistics, a sample is the set of all those observations actually recorded. A sample is a subset of a parent **population**. (b) In geology, a sample is each of the observations or specimens in a survey.

SAMPLE DESCRIPTION SPACE. An alternative term for **sample space**.

SAMPLE FUNCTION. A synonym for **realization**.

SAMPLE SIZE. In a statistical survey, the total number of observations.

SAMPLE SPACE. The sample space Ω of a random phenomenon is the set of all possible **outcomes** of the phenomenon.

SCALAR PRODUCT. An alternative term for **inner product**.

SCATTERGRAM. A diagram in which values of one variable are plotted against corresponding values of another variable primarily for the purpose of visualizing potential relationships among the variables. A scattergram appears as a cloud of points. If the cloud is concentrated along a smooth curve, then there is a better potential to **model** the relationship through **regression**.

SCATTERPLOT. A synonym for **scattergram**.

SCHWARZ INEQUALITY. If $\boldsymbol{x}$ and $\boldsymbol{y}$ are **vectors** in an **inner product space**, then

$$|\langle \boldsymbol{x},\boldsymbol{y}\rangle| \leq \|\boldsymbol{x}\|\,\|\boldsymbol{y}\|$$

This inequality implies that $-1 \leq \langle \boldsymbol{x},\boldsymbol{y}\rangle / \|\boldsymbol{x}\|\,\|\boldsymbol{y}\| \leq 1$, so the quotient $\langle \boldsymbol{x},\boldsymbol{y}\rangle / \|\boldsymbol{x}\|\,\|\boldsymbol{y}\|$ is the cosine of exactly one angle between 0 and 180 degrees, which in a **euclidean space** of two or three dimensions is the geometric angle between the vectors. For two real **functions** $f(x)$ and $g(x)$, the inequality becomes

$$\left|\int_a^b f(x)g(x)dx\right|^2 \leq \left|\int_a^b f^2(x)dx\right|\left|\int_a^b g^2(x)dx\right|$$

which is also known as the *Cauchy–Schwarz inequality.*

SCREEN EFFECT. In **kriging**, the reduction of the influence of an observation in the **estimation** of another, caused by the addition of one or more **sample** points at intermediate locations between the original observation and the estimation location. The screen effect decreases as the **nugget effect** increases. Inverse distance estimation procedures do not take into account the screen effect.

SECOND ORDER RANDOM FIELD. A **random field** is second order stationary if it has **stationarity of order 2.**

SECOND ORDER RANDOM VARIABLE. A **random variable** whose **moments**, up to the second order, are finite.

SELECTIVE. Given two **probability density functions** $f_1(x)$ and $f_2(x)$, $f_1(x)$ is less selective than $f_2(x)$ if and only if

$$\int_0^\infty \varphi(z)f_1(z)dz \leq \int_0^\infty \varphi(z)f_2(z)dz$$

for any **convex function** $\varphi(z)$. It can be proven that both probability density functions have the same **mean**, and that the **variance** of $f_1(x)$ is less than or equal to the variance of $f_2(x)$.

SELECTIVITY CURVE. Let $F(x)$ be a **distribution function** used to characterize the grade of metal in an ore body. Let the **random function** used to **model** the **regionalized variable** be **strictly stationary**. Let z be the cutoff grade. Four **functions**, collectively called selectivity curves, have been defined to study the fluctuations in ore grades. The tonnage-grade curve is defined as

$$T(z) = 1 - F(z)$$

The metal quantity-grade curve is

$$Q(z) = \int_z^\infty u\,F(du)$$

From these two curves, by means of a change in variables $t = T(z)$, the quantity of metal as a function of tonnage is obtained

$$Q(t) = Q(z)$$

Finally, the conventional benefit curve is defined by

$$B(z) = Q(z) - zT(z)$$

The properties of all four curves are strongly related to the **selective** nature of the distribution function $F(x)$. All of these functions are used for economic evaluation in mining.

SELFSIMILAR. A geometrical shape is said to be selfsimilar if its structure is such that magnification by a given factor reproduces the shape of the original object, hence it is not possible to determine the scale of the geometric shape from its form. **Isotropic** forms of **fractals** are selfsimilar.

SEMIMARKOVIAN RANDOM FUNCTION. Let C, K, and L be **compact** sets in an n-dimensional **euclidean space** such that any segment from a point in K to a point in L intersects C. A **random function** $\boldsymbol{Z(x)}$ in C is said to be semimarkovian if, given $\boldsymbol{Z(x)}$, there is **stochastic independence** between **events** in K and L. This **model** is used in the **simulation** of **random variables**.

SEMIVARIOGRAM. (a) Let $Z(\cdot)$ be a **random function** modeling a **regionalized variable** $z(\cdot)$. Given two locations $\boldsymbol{x}$ and $(\boldsymbol{x}+\boldsymbol{h})$, a semivariogram is a **measure** of one half the **mean square error** produced by assigning the value $z(\boldsymbol{x}+\boldsymbol{h})$ to the value $z(\boldsymbol{x})$. In a **probabilistic** sense, the semivariogram, $\gamma(\boldsymbol{h})$, is one half of the **variogram of order 2**

$$\gamma(\boldsymbol{h}) = 0.5\,\mathrm{Var}\{Z(\boldsymbol{x}) - Z(\boldsymbol{x}+\boldsymbol{h})\} = 0.5\,\mathrm{E}[\{Z(\boldsymbol{x}) - Z(\boldsymbol{x}+\boldsymbol{h})\}^2]$$

Because the increments are a **function** of the distance between pairs of locations, the semivariogram is a function of **lag** rather than a single numerical value. For random functions that are second order **stationary**, the semivariogram is related to the **covariance** $\mathrm{Cov}(\cdot)$ through the formula

$$\gamma(\boldsymbol{h}) = \mathrm{Cov}(0) - \mathrm{Cov}(\boldsymbol{h})$$

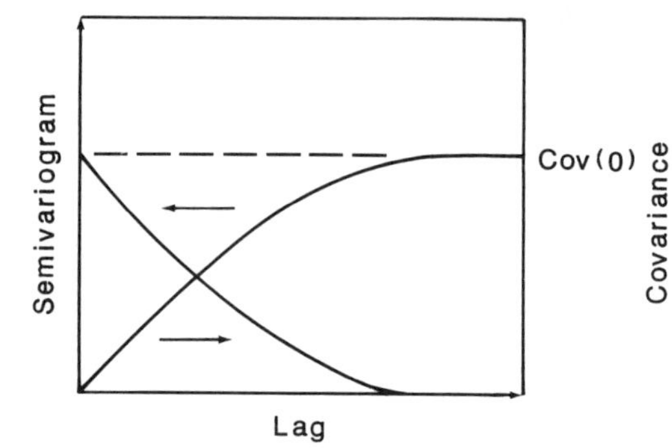

Traditionally, estimating the semivariogram is preferable to estimating the covariance because the **experimental semivariogram** does not require knowledge or a prior **estimate** of the **mean** of the **population**. To assure that **permissibility criteria** are honored, a tabulation of the experimental semivariogram is never used in the **models** available for **prediction**, but rather one of the permissible **semivariogram models** that best represents the experimental semivariogram. Some authors like to drop the prefix semi- and refer to the semivariogram simply as the **variogram**.

SEMIVARIOGRAM FITTING. The selection of analytical expressions and parameters for a **semivariogram model** of an **experimental semivariogram**. There are four hidden steps in this process. The **semivariogram** to be **modeled** is the **theoretical semivariogram**, but it is not known and cannot be calculated because the theoretical semivariogram is not a **regional characteristic**. Provided the **intrinsic hypothesis** holds, it is possible to replace the theoretical semivariogram by the **expected value** of the **probabilistic regional semivariogram**, but since only one **realization** exists instead of the infinite number of realizations demanded by the definition of expectation, this expectation cannot be known. The **regional semivariogram** could be **estimated** using only the single realization, but since an exhausted sampling is required, the best that can be done with a finite sampling is to calculate a representative experimental semivariogram. This is the best alternative and the reason why the **curve fitting** is done to the experimental semivariogram rather than to the theoretical semivariogram.

SEMIVARIOGRAM MODEL. A mathematical expression that results from **curve fitting** a **function** to the **experimental semivariogram**. The analytical expression must satisfy the **permissibility criteria**; possibilities include the **cubic semivariogram**, **exponential semivariogram**, **Gaussian semivariogram**, **linear semivariogram**, **nested semivariogram**, **penta-spherical semivariogram**, **pure nugget effect semivariogram model**, or **spherical semivariogram**.

SEQUENTIAL INDICATOR SIMULATION. A procedure used for the numerical **simulation** of a continuous **random function** in an n-dimensional **euclidean space**. The technique allows the use of **hard data, soft data,** or both. The algorithm generates a sequence of **distribution functions** based on the original **data** as well as on all previously simulated values. These conditional distribution functions are built using either **indicator kriging, probability kriging,** or **multigaussian kriging** when the random function follows a **multinormal distribution**. Sequential indicator kriging is designed to study random functions that clearly depart from a multinormal distribution, and has been used to produce simulated images of spatial connectivity in porous media and to assess the **uncertainty** of the characterization.

SERIAL VARIATION FUNCTION. An alternative term for **semivariogram**.

SICHEL t DISTRIBUTION. A **distribution** for calculating **confidence limits** around the **mean** of a multi**lognormal distribution**. The Sichel t distribution is available in tabular form for various **probability** levels, **sample sizes**, and natural logarithmic **variances**.

SIGMA-FIELD. A synonym for **σ-algebra**.

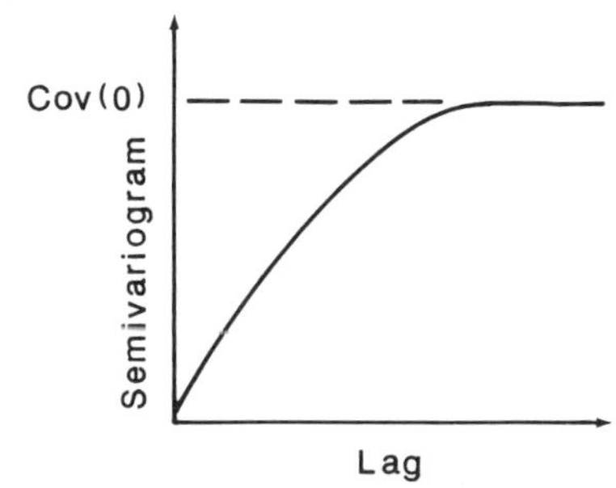

SILL. The limiting value for large arguments of a **transitive** type of **semivariogram**. The sill is numerically equal to the **a priori variance** Cov(0).

SIMILARITY DIMENSION. A synonym for **Hausdorff–Besicovitch dimension**.

SIMPLE KRIGING. Generalized linear **regression** under the assumption of second order **stationarity** with a known **mean**. The **normal equations** are the same as those for **ordinary kriging** except that the final row is deleted from all matrices, as is the final **column** of the **square matrix**. The **semivariogram** $\gamma(\cdot)$ or the **covariance** Cov($\cdot$) can be used interchangeably, taking advantage of the equivalence

$$\gamma(\boldsymbol{h}) = \text{Cov}(0) - \text{Cov}(\boldsymbol{h})$$

Simple kriging is the best of all linear or nonlinear **minimum mean square error estimators** only when a **multinormal distribution** is used as a **model** for the **random function**. The method can be regarded as a version of the **Wiener filter**.

SIMULATION. The use of **models** to study the nature of certain phenomena. In **geostatistics** the usage is restricted to synthetic **realizations** of a **random function** that possess the same spatial **statistics** as the actual realization, as measured by **moments** up to order 2. While a **kriging** surface is unique, there are an infinite number of simulated surfaces for the same **data** and statistical moments. **Conditional simulation** honors the data, but unconditional simulation does not. Because of the impossibility of **estimating** an actual surface, simulation is preferred over kriging when a model of the spatial fluctuations is more important than **accuracy** at the **local** scale. For a given **sample**, at any unsampled location the conditional simulation **estimation variance** is larger than the **kriging variance** at the same location. Procedures used in the simulation of **regionalized variables** include the **turning bands method, lower-upper (LU) decomposition, random tokens methods, sequential indicator simulation** and spectral techniques.

SINGLE INDICATOR KRIGING. An approximation of **simple kriging** or **ordinary kriging** of **indicator data columns**. The approximation is made by assuming that all indicator **semivariograms** at all threshold values are proportional to each other. The approximation is justified in the presence of a **mosaic model**. This approach is also called *median indicator kriging* because, when the mosaic model holds, the single **indicator** semivariogram that is required can best be **estimated** by the **experimental semivariogram** of the indicator **transforms** provided by the **median** threshold value. This approximation, although practical, is seldom used in practice.

SINGULAR MATRIX. A **square matrix** with no **inverse matrix**. The **determinant** of a singular matrix is 0.

SKEWNESS. Departure of a **probability density function** from symmetry. The **coefficient of skewness** is the most frequently used **measure** of this property.

SLIDING NEIGHBORHOOD. Given a **neighborhood** V around a point $\boldsymbol{x}$, a sliding neighborhood for any point $\boldsymbol{y}$ is the new set of points that results from translating V by a **vector** $\boldsymbol{y}-\boldsymbol{x}$.

SLIDING REPRESENTATION. Let V be a **sliding neighborhood** and let S be a subset of a **euclidean space**. Let $z(\cdot)$ be a **regionalized variable**. The sliding representation of $z(\cdot)$ in S for the sliding neighborhood V is the **random function** $Z(\cdot)$

$$Z(\boldsymbol{x}) = z(\boldsymbol{x} + \overline{\boldsymbol{y}})$$

where $\boldsymbol{x}$ is any point in V and $\overline{\boldsymbol{y}}$ and an equally probable and randomly located point in S. This representation allows **stationarity** to be defined as a characteristic of the **estimator** rather than an intuitive property of the regionalized variable. Sliding representation provides a link between **transitive methods** and **topoprobabilistic models**.

SMOOTHING. The process of removing short scale or high frequency fluctuations in a **time series** or **regionalized variable** so that the new set of values has more regular first differences and smaller higher order differences. Although smoothing can be carried out graphically, it is commonly done by **moving linear averages** or **least squares methods** of **curve fitting**.

SOFT DATA. (a) Information that lacks **accuracy** and is used to supplement insufficient **hard data**. An **indicator data column** associated with inaccurate information is defined only for threshold values outside a **constraint interval**. Soft data are used together with hard data in **indicator kriging** and **indicator cokriging**. (b) A **local** form of **fuzzy indicator data** for an unknown value $z(\boldsymbol{x})$.

SPACE TRANSFORMATION OF THE FIRST KIND. Let k be an integer number larger than 0 and let $\{H_{n-k}; \boldsymbol{s}, \boldsymbol{x}\cdot\boldsymbol{s}\}$ be a set of lines passing through a given point in the n-dimensional **euclidean space** $\Re^n$. The subscript $n-k$ denotes the **codimension** of the lines H_{n-k} considered as hyperplanes in $\Re^n$, and $\boldsymbol{s}$ is a unit **vector** defining the orientation of H_{n-k}. The **inner product** $\boldsymbol{x}\cdot\boldsymbol{s}$ defines the distance of the lines from the origin so that

$$\int_{\Re^n} \delta_t(\boldsymbol{x}\cdot\boldsymbol{s})\,d\boldsymbol{x} = \begin{cases} 1, & \text{if } t = \boldsymbol{x}\cdot\boldsymbol{s} \\ 0, & \text{if } t \neq \boldsymbol{x}\cdot\boldsymbol{s} \end{cases}$$

A space transformation of the first kind, T_n^k, assigns to a **function** $f_n(\boldsymbol{x})$ in $\Re^n$ the following set of functions

$$\mathrm{T}_n^k[f_n(\boldsymbol{x})] = \int_{\Re^n} f_n(\boldsymbol{x})\,\delta_t(\boldsymbol{x}\cdot\boldsymbol{s})\,d\boldsymbol{x}$$

on the lines $\{H_{n-k}; \boldsymbol{s}, \boldsymbol{x}\cdot\boldsymbol{s}\}$. Space transformations of the first kind and the **space transformation of the second kind** share dual characteristics. In general, space transformations of the first kind integrate over hyperplanes of codimension $n-k$, and space transformations of the second kind integrate over the set of hyperplanes passing through a point in $\Re^n$.

SPACE TRANSFORMATION OF THE SECOND KIND. Let $f_{k,s}(\boldsymbol{x}\cdot\boldsymbol{s})$ be a set of **functions** defined over a set of lines $\{H_{n-k};\boldsymbol{s},\boldsymbol{x}\cdot\boldsymbol{s}\}$ passing through a given point in the n-dimensional **euclidean space** $\Re^n$. A space transformation of the second kind, Ψ_k^n of a set of functions $f_{k,s}(\boldsymbol{x}\cdot\boldsymbol{s})$, is the **operator**

$$\Psi_k^n[f_{k,s}(\boldsymbol{x}\cdot\boldsymbol{s})] = \int_{\Theta_n} u(\boldsymbol{x},\boldsymbol{s}) f_{k,s}(\boldsymbol{x}\cdot\boldsymbol{s})d\boldsymbol{s}$$

where $\boldsymbol{s}$ is a unit **vector** defining the orientation of the hyperplane, and $u(\boldsymbol{x},\boldsymbol{s})$ is a weight function. The integration is carried out over the **closed** surface Θ_n in $\Re^n$. The **turning bands method** is a special case of this space transformation of the second kind.

SPATIAL DEPENDENCE. The set of relations linking the values of a **regionalized variable** at different locations in space. Spatial dependence usually arises from a common genesis, such as a certain mineralization process in the case of an ore grade, and is used with advantage in **estimating** an unsampled value from neighboring observations. In circumstances where the regionalized variable is regarded as a **realization** of a **random function**, spatial dependence is **modeled** by **stochastic dependence**.

SPATIAL INDEPENDENCE. Lacking **spatial dependence**.

SPEARMAN CORRELATION COEFFICIENT. A **correlation coefficient** between the **rank order transformation** of two **random variables** or two sets of observations having the same **sample size**. If the Spearman correlation coefficient is 1, the ranks of the two variables are identical, but the relationship between the variables may not be linear. Unlike the **Pearson correlation coefficient**, the Spearman correlation is not strongly influenced by **outliers**. Significant differences between the Pearson and the Spearman correlation coefficients often indicate either a nonlinear relationship or the presence of extreme values in at least one of the sets of measurements. If the two sets of observations are not the same size, their rank orders can be standardized by dividing by their respective sample sizes, defining two sets of **uniform transforms**. The correlation coefficient between these two sets of uniform transforms has the same significance as a Spearman correlation coefficient.

SPHERICAL SEMIVARIOGRAM. The **transitive** form of **semivariogram model** defined by

$$\gamma(h) = \begin{cases} C\left[\dfrac{3}{2}\left(\dfrac{h}{a}\right) - \dfrac{1}{2}\left(\dfrac{h}{a}\right)^3\right] & , \ 0 \le h < a \\ C & , \ a \le h \end{cases}$$

where C is the **sill** and a is the **range**. Near the origin this semivariogram is close to a line in shape. The tangent to this semivariogram at the origin intersects the sill at 2/3 the range. In practice, this is one of the models that is most commonly fitted to **experimental semivariograms** in **euclidean spaces** of dimensions not greater than 3.

SPLINE FUNCTION. A mathematical counterpart of the flexible ruler used in drafting to draw a curve through a sequence of points. Let $x_1, x_2, \ldots, x_i, \ldots x_k$ be a strictly increasing sequence of abscissas on the plane and let $[x_0, x_{k+1}]$ be an interval enclosing the sequence. A spline function, $S(x)$, of degree n for the sequence $x_1, x_2, \ldots, x_i, \ldots x_k$ is a **function** having the following properties

i. In each interval $[x_i, x_{i+1}]$, for $i = 0, 1, \ldots, k+1$, $S(x)$ is given by a **polynomial of degree** n or less.

ii. $S(x)$ and its derivatives of order less than or equal to $(n-1)$ are continuous everywhere.

Spline functions are **exact interpolators**. The concept has been generalized to interpolate surfaces in three dimensions. Physically, for $n=3$, cubic spline functions correspond to a surface having minimum tension. **Curve fitting** of spline functions is equivalent to **kriging** with a specific **model** for the **covariance**.

SQUARE MATRIX. A **matrix** in which the number of **rows in the matrix** is equal to the number of **columns in the matrix**.

STANDARD DEVIATION. The positive square root of the **variance**. The standard deviation has the same units of measurement as the variable from which it is calculated.

STANDARD ERROR. The **standard deviation** of the difference between an actual parameter and its **estimate** for an **estimator** without **bias**.

STANDARD NORMAL DISTRIBUTION. Let X be a **random variable** that follows a **normal distribution** with **mean** μ and **variance** σ^2, and let Y be the linear **transformation**

$$Y = \frac{X - \mu}{\sigma}$$

The **variate** Y is said to follow a standard normal distribution, which is simply a normal distribution with a mean of 0 and a variance of 1.

STANDARD SCORE. A synonym for **z-score**.

STATIONARITY. A term used to denote different degrees of invariance in the characteristics of **random functions**. Stationarity refers to the **model** and not to the **data**. See also **stationary in the wide sense, stationary of finite order, strict stationarity**, and **weak stationarity**.

STATIONARY. The quality of having **stationarity**.

STATIONARY IN THE WIDE SENSE. Displaying **weak stationarity**.

STATIONARY OF FINITE ORDER. A **stochastic process** or **random function** is order n stationary when the **joint probability distribution** of any number up to a number n of component **random variables** is invariant under any joint translation of these components in space. Second order stationarity, in addition to the invariance by translation of any **bivariate distribution**, is customarily understood to imply that the **covariance** exists and depends only on the distance $\boldsymbol{h}$ between pairs of locations and not on their individual spatial locations. This further implies that both the **variance** and the **mean** exist and do not depend on spatial location. **Strict stationarity** does not imply second order stationarity.

STATISTIC. A **function** of the observations comprising a **sample**. A statistic is usually, but not necessarily, an **estimator** of some parameter of a **distribution function**.

STATISTICAL DEPENDENCE. A synonymous term for **stochastic dependence**.

STATISTICAL INDEPENDENCE. An alternative term for **stochastic independence**.

STEP FUNCTION. A **function** $f(x)$ defined over the interval (a,b) is a step function if (a,b) can be divided into a finite number of subintervals over each of which $f(x)$ is a constant. A **Poisson process** is a step function.

STOCHASTIC DEPENDENCE. The condition of not possessing **stochastic independence**.

STOCHASTIC INDEPENDENCE. Lack of relationship between **events** or **random variables**. Events $\{\mathcal{A}_1, \mathcal{A}_2, \ldots, \mathcal{A}_i, \ldots, \mathcal{A}_n\}$ are independent if the occurrence or nonoccurrence of one event does not change the **probability** of occurrence of any of the other events. The events $\{\mathcal{A}_1, \mathcal{A}_2, \ldots, \mathcal{A}_i, \ldots, \mathcal{A}_n\}$ are stochastically independent if and only if

$$P\left(\bigcap_{i=1}^{n} \mathcal{A}_i\right) = \prod_{i=1}^{n} P(\mathcal{A}_i)$$

Random variables $\{X_1, X_2, \ldots, X_i, \ldots, X_n\}$ are stochastically independent if and only if

$$P(X_1 \le x_1, X_2 \le x_2, \ldots, X_i \le x_i, \ldots, X_n \le x_n) = \prod_{i=1}^{n} P(X_i \le x_i)$$

for any choice of numbers x_i. The term is a synonym for *statistical independence*.

STOCHASTIC PROCESS. A collection of **random variables** $\{X(t,\omega): t \in T, \omega \in \Omega\}$, where T is an index set, and Ω is the **sample space**. There are as many random variables $X(t,\omega)$ as there are elements in T, one for each t. The term often is applied when t varies continuously within a one-dimensional **euclidean space** $\Re^1$, such as time. The term **random function** is used when t, usually denoted by $\boldsymbol{x}$, varies continuously in a euclidean space of a higher dimension.

STOCHASTIC VARIABLE. A synonym for **random variable.**

STRATIFIED ANISOTROPY. An alternative term for **zonal anisotropy.**

STRATIFIED RANDOM SAMPLING. A **sample** in which observations are drawn at random from sub**populations**, called *strata*. The subpopulations have a regular, ordered relation to one another.

STRICTLY CONDITIONALLY POSITIVE DEFINITE FUNCTION. A **conditionally positive definite function** for which, if the equality holds, all coefficients α_i are equal to 0.

STRICTLY POSITIVE DEFINITE FUNCTION. A **positive definite function** for which, if the equality holds, all coefficients α_i are equal to 0.

STRICT STATIONARITY. A **stochastic process** is stationary in the strict sense if its **statistics** are invariant under a shift in the origin. From the definition it follows that any two k-component vectorial processes $\{Z(x_1), Z(x_2), \ldots, Z(x_k)\}$ and $\{Z(x_1+h), Z(x_2+h), \ldots, Z(x_k+h)\}$ have the same **joint distribution function**, regardless of the number k of components and the distance h. Observations can be regarded as coming from a single **probability density function** rather than coming from individual **distributions** at each location. Strict stationarity does not imply that any particular **moments**, in particular the **mean** and **covariance**, exist or are finite. However, if these moments do exist, then they are stationary. This form of **stationarity** is too restrictive in many circumstances. **Geostatistics** is based on less restrictive forms of stationarity such as **weak stationarity** and the **intrinsic hypothesis.**

STRONG STATIONARITY. An alternative term for **strict stationarity.**

STRUCTURAL ANALYSIS. Inspection and numerical **modeling** of the patterns of **spatial dependence** that characterize a **regionalized variable.** If a **random function** model is considered, structural analysis involves decisions about **stationarity** and inference as well as the modeling of **distribution functions** and **structural functions.**

STRUCTURAL FUNCTION. A generic name for any of the **functions**, such as the **semivariogram** or the **covariance**, that **model** the pattern of **spatial dependence** of a **regionalized variable.**

STRUCTURE. An attribute denoting the possession of spatial **correlation. Regionalized variables** are characterized by having both random and structured fluctuations that are better **modeled** by

random functions than by **stochastically independent** forms of **random variables** as used in classical statistics.

STRUCTURE FUNCTION. An alternative term for **semivariogram**.

SUPPORT. (a) An n-dimensional volume within which linear **average values** of a **regionalized variable** may be computed. The complete specification of the support includes the geometrical shape, size, and orientation of the volume. The support can be as small as a point or as large as the entire **field**. A change in any characteristic of the support defines a new regionalized variable. Changes in the regionalized variable resulting from alterations in the support can sometimes be related analytically. (b) The support of a **function** is the set where the function is not identically equal to 0.

SUPPORT EFFECT. The influence that the size of the **support** has on **average values** of the **regionalized variable**. Although **globally** the **mean** remains unchanged, in general, a larger support will have a smaller **dispersion variance** within any **field** and, in particular, a smaller **a priori variance** and more symmetric **distribution**. The increase in symmetry is a manifestation of the **central limit theorem**. For example, in a **selective** mining operation, a larger size of mining unit will result in smaller fluctuations in average grades.

SYMMETRIC MATRIX. A **matrix** that is equal to the **transpose of the matrix**, such as

$$S = S^T = \begin{bmatrix} 16 & 74 & 0 \\ 74 & 98 & 23 \\ 0 & 23 & 0.1 \end{bmatrix}$$

THEORETICAL SEMIVARIOGRAM. A characteristic of a **topoprobabilistic model** that, for short, is usually referred to as a **semivariogram**. The theoretical semivariogram is not a **regional characteristic**.

THEORY. A systematic account of some field of study derived from a set of general propositions. In pure mathematics, these propositions tend to be axiomatic, as in the theory of numbers, while in the natural sciences the propositions tend to be regarded as more primitive, less in need of justification, and to have some **experimental** background, as in the theory of relativity or continental drift. There may be rival theories in a particular field, differing in the selection or the emphasis laid on the principles, such as the different theories about the propagation of light.

THIESSEN POLYGON. An alternative term for **Voronoi polygon**.

THREE-PARAMETER LOGNORMAL DISTRIBUTION. A **random variable** X follows a three-parameter lognormal distribution if its natural logarithmic **transformation** Y

$$Y = \log_e(X + \beta)$$

follows a **normal distribution**. The distribution is uniquely defined by the **mean**, **variance** and shift parameter β.

TIME SERIES. A sequence, discrete or continuous, of **data** assigned to specific instants in time and studied with respect to the **statistics** of the **distribution** in time, such as a record of water table elevations from an observation well.

TOPOPROBABILISTIC MODEL. Any approach using the notion of **probability** to describe a phenomenon varying in space. In **geostatistics**, the representation of a **regionalized variable** by a

random function is a topoprobabilistic model.

TRANSFER FUNCTION. (a) A **function** that **transforms** an input sequence into a response function. (b) In **geostatistics**, an obsolete synonym for **recovery function**.

TRANSFORM. (a) The value of a **transformation** at a given point. (b) Given a **function** $F(\cdot)$, its transform $f(x)$ is the integral

$$f(x)=\int_a^b K(t,x)F(t)\,dt$$

where $K(t,x)$ is a known function. The limits of integration may be either constants or functions of x.

TRANSFORMATION. The term transformation is essentially a synonym of such words as **function** or **operator** and is used to denote a passage from one figure or expression to another. Most transformations are performed to simplify calculations or to meet assumptions.

TRANSITION PROBABILITY. Given a **stochastic process** $X(t)$, the transition probability between two times t_0 and t_1 is the **conditional probability**

$$P[X(t_1)\leq \tau_1 | X(t_0)=\tau_0]$$

TRANSITIVE. In a **semivariogram**, the quality of having a finite **sill** equal to the **a priori variance**. The sill is reached at a finite distance called the actual **range**, or it is approached asymptotically, in which case there is a practical range given by the distance at which the **semivariogram model** is equal to 95 percent of the value of the sill. All transitive semivariogram models $\gamma(\boldsymbol{h})$ have a **covariance** counterpart $\mathrm{Cov}(\boldsymbol{h})$ given by

$$\mathrm{Cov}(\boldsymbol{h})=\mathrm{Cov}(0)-\gamma(\boldsymbol{h})$$

TRANSITIVE COVARIOGRAM. Let $f(\boldsymbol{x})$ be a **regionalized variable** that vanishes outside a bounded **field** V. The transitive covariogram of this regionalized variable is the **function** $g(\boldsymbol{x})$ defined by

$$g(\boldsymbol{h})=\int_V f(\boldsymbol{x})f(\boldsymbol{x}+\boldsymbol{h})\,d\boldsymbol{x}$$

The transitive covariogram has the property that $g(\boldsymbol{h})\leq g(0)$ and is also symmetric, that is, $g(\boldsymbol{h})=g(-\boldsymbol{h})$.

TRANSITIVE KRIGING. A **transitive method** of forecasting values of a **regionalized variable**. The **normal equations** are similar to those used in **simple kriging**, except that **transitive covariograms** are used instead of **covariances**. This form of **kriging** shares all the advantages and disadvantages of transitive methods. On the positive side, there is no assumption of **stationarity**, which makes transitive kriging attractive for use where a **trend** is present. Drawbacks include difficulties in calculating a transitive covariance **model** and computational problems that make it difficult to **estimate** the regionalized variable near the sampling borders.

TRANSITIVE METHOD. Any of a collection of procedures for studying **regionalized variables** that are based on geometric considerations rather than notions of **probability** or **stationarity**. Transitive results are strongly influenced by edge effects, which is partly why these methods are used primarily to solve geometrical problems. In theory, any regionalized variable that has a finite **support** can be analyzed by transitive methods. By and large, a transitive approach is more general than a **topoprobabilistic model** approach but is more analytically complex. The operation of **grading** is a notable exception. Although the computations in any transitive method are purely deterministic, some results may be interpreted only by incorporating the notion of probability. Transitive methods are different from, but related to, **deterministic geostatistics**.

TRANSLATION OF PERMISSIBLE COEFFICIENTS. The translation by a **vector** $\boldsymbol{h}$ of **permissible coefficients** $\{\alpha_1, \alpha_2, \ldots, \alpha_i, \ldots, \alpha_n\}$, each one linked to a location $\boldsymbol{x}_i$, is the set of permissible coefficients $\{\alpha_1, \alpha_2, \ldots, \alpha_i, \ldots, \alpha_n\}$ linked to locations $\boldsymbol{x}_i + \boldsymbol{h}$.

TRANSPOSE OF A MATRIX. The transpose $\boldsymbol{A}^T$ of a **matrix** $\boldsymbol{A}$ is defined as the matrix that results from interchanging the rows and columns of matrix $\boldsymbol{A}$.

TREND. The trend is an abstract expression of the low frequency, large-scale variation of a **regionalized variable**. The trend is often an intuitive concept that is made equivalent to quantitative notions such as the **moving linear average** or the **convolution** of a regionalized variable. A trend can also be **modeled** by the **drift**. Since the trend is relative to the **working scale**, a regionalized variable does not possess a unique trend. The same feature may be regarded as part of the trend at a **local** level, as an anomaly at a regional level, and as an insignificant detail at the **global** scale.

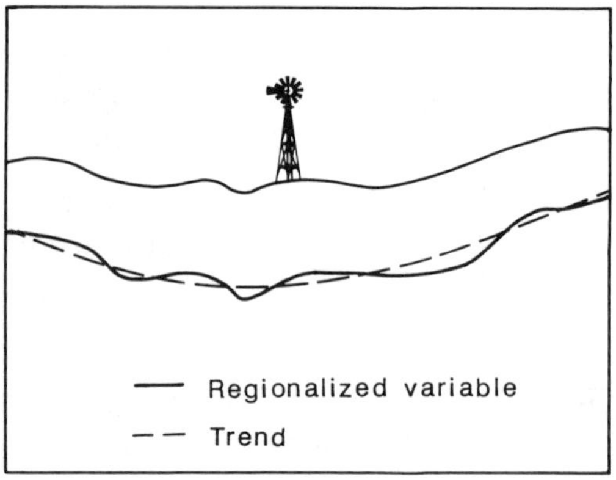

TRIAL. A single performance of a **random experiment**.

TURNING BANDS METHOD. A special case of a **space transformation of the second kind** used for numerical **simulation** of n-dimensional **realizations** of a **random function**. The n-dimensional realizations are obtained by adding the contributions of several independent one-dimensional realizations along lines that are randomly or **uniformly distributed** in a **euclidean space**. Because of the **central limit theorem**, the method will produce realizations that are **multinormally distributed**. The turning bands method does not directly produce a **conditional simulation**.

TWO-PARAMETER LOGNORMAL DISTRIBUTION. A synonym for **lognormal distribution**. The two-parameter denomination refers to the logarithmic **mean** and **variance**, the only parameters required to completely determine a lognormal distribution.

UNBIASED. Without **bias**.

UNCERTAINTY. Lack of assurance about the truth of a statement or about the exact magnitude of an unknown measurement or number. Uncertainty is a central concept in the decision making that usually follows any **geostatistical** study, in which case it is related to the risk of making an incorrect decision because the **estimates** do not agree with reality. Uncertainty is often measured by parameters such as **confidence intervals**, **probability intervals**, and **standard errors**.

UNDERESTIMATION. See **error**.

UNDERLYING SEMIVARIOGRAM. The **semivariogram** for the **residuals** of a **regionalized variable** being **modeled** by a **random function** that is not **stationary**.

UNIFORM CONDITIONING. A family of approximation methods that replaces the conditioning of several **random variables** by a single **function** of the variables. The theoretical justification for these methods is the following theorem. Let $\{Y_0, Y_1, \ldots, Y_i, \ldots, Y_n\}$ be a collection of random variables that follow a **multinormal distribution.** Let $f(Y_0)$ be any function of Y_0 and let Y_0^* be the **simple kriging estimator** of Y_0 by the remaining random variables. Then,

$$\mathrm{E}[f(Y_0)|Y_1,\ldots,Y_i,\ldots,Y_n] = \mathrm{E}[f(Y_0)|Y_0^*]$$

Uniform conditioning may be used to **estimate** a **recovery function** when the **model** for the **random function** is a multinormal distribution.

UNIFORM DISTRIBUTION. The **random variable** X has a uniform distribution over the interval $[a,b]$ if its **probability density function** $f(x)$ is given by

$$f(x) = \begin{cases} \dfrac{1}{b-a} & , \; a \le x \le b \\ 0 & , \; \text{elsewhere} \end{cases}$$

The **mean** of the uniform distribution is $0.5(a+b)$ and its **variance** is $(b-a)^2/12$.

UNIFORM TRANSFORMATION. A **transformation** of a **random function** $Z(\cdot)$ that is **stationary** with a **marginal distribution** $F(z)$ into another stationary random function $U(\cdot)$ having a marginal distribution that is **uniformly distributed** over the interval [0,1]. The transformation is the **distribution function** $F(z)$ of $Z(\cdot)$.

$$U(\boldsymbol{x}) = F[Z(\boldsymbol{x})]$$

If $Z(\boldsymbol{x})$ is a **discrete random variable** with n possible **outcomes**, then $nU(\boldsymbol{x})$ is the **rank order transform** of $Z(\boldsymbol{x})$. Uniform transforms are used in the **Monte Carlo method** and in **probability kriging.**

UNIMODAL DISTRIBUTION. A **distribution** having a single **mode.**

UNION STABILITY. A property of a **random set** S that is **closed** so that for any number n, it is possible to find a collection of closed random sets S_i that are mutually **stochastically independent,** have the same **Choquet's capacity,** and have a positive real number λ_n

$$\lambda_n \bigcup_{i=1}^{n} S_i$$

that has the same Choquet's capacity as $\lambda_n S$. Such a random set S is **selfsimilar.**

UNIVARIATE. Comprising a single variable.

UNIVERSAL. Honoring the **universality condition.**

UNIVERSALITY CONDITION. In **universal kriging,** any of the requirements that state that an **estimation of error** should be **unbiased,** whatever the unknown **drift** might be. In addition to being an unbiasedness constraint, the purpose of the universality conditions is to **filter** out the unknown drift components in the **random function.**

UNIVERSAL KRIGING. Simple kriging of the **residuals** of a **regionalized variable** after automatically removing optimal **estimates** of the **drift.** Provided that the drift can be modeled within **local neighborhoods** by

$$\sum_{j=0}^{m} a_j\, p^j(\boldsymbol{x})$$

the simple kriging system of equations expands to a constrained system of **normal equations**

$$X = A^{-1}B$$

where

$$\boldsymbol{X} = \begin{bmatrix} \lambda_1 \\ \cdots \\ \lambda_i \\ \cdots \\ \lambda_n \\ \mu_0 \\ \mu_1 \\ \cdots \\ \mu_j \\ \cdots \\ \mu_m \end{bmatrix} \qquad \boldsymbol{B} = \begin{bmatrix} \gamma(\boldsymbol{x}_1,\boldsymbol{x}_0) \\ \cdots \\ \gamma(\boldsymbol{x}_i,\boldsymbol{x}_0) \\ \cdots \\ \gamma(\boldsymbol{x}_n,\boldsymbol{x}_0) \\ 1 \\ p^1(\boldsymbol{x}_0) \\ \cdots \\ p^j(\boldsymbol{x}_0) \\ \cdots \\ p^m(\boldsymbol{x}_0) \end{bmatrix}$$

$$\boldsymbol{A} = \begin{bmatrix}
\gamma(0) & \cdots & \gamma(\boldsymbol{x}_1,\boldsymbol{x}_i) & \cdots & \gamma(\boldsymbol{x}_1,\boldsymbol{x}_n) & 1 & p^1(\boldsymbol{x}_1) & \cdots & p^j(\boldsymbol{x}_1) & \cdots & p^m(\boldsymbol{x}_1) \\
\cdot & \cdot & \cdot & \cdot & \cdot & \cdot & \cdot & \cdot & \cdot & \cdot & \cdot \\
\gamma(\boldsymbol{x}_i,\boldsymbol{x}_1) & \cdots & \gamma(0) & \cdots & \gamma(\boldsymbol{x}_i,\boldsymbol{x}_n) & 1 & p^1(\boldsymbol{x}_i) & \cdots & p^j(\boldsymbol{x}_i) & \cdots & p^m(\boldsymbol{x}_i) \\
\cdot & \cdot & \cdot & \cdot & \cdot & \cdot & \cdot & \cdot & \cdot & \cdot & \cdot \\
\gamma(\boldsymbol{x}_n,\boldsymbol{x}_1) & \cdots & \gamma(\boldsymbol{x}_n,\boldsymbol{x}_i) & \cdots & \gamma(0) & 1 & p^1(\boldsymbol{x}_n) & \cdots & p^j(\boldsymbol{x}_n) & \cdots & p^m(\boldsymbol{x}_n) \\
1 & \cdots & 1 & \cdots & 1 & 0 & 0 & \cdots & 0 & \cdots & 0 \\
p^1(\boldsymbol{x}_1) & \cdots & p^1(\boldsymbol{x}_i) & \cdots & p^1(\boldsymbol{x}_n) & 0 & 0 & \cdots & 0 & \cdots & 0 \\
\cdot & \cdot & \cdot & \cdot & \cdot & \cdot & \cdot & \cdot & \cdot & \cdot & \cdot \\
p^j(\boldsymbol{x}_1) & \cdots & p^j(\boldsymbol{x}_i) & \cdots & p^j(\boldsymbol{x}_n) & 0 & 0 & \cdots & 0 & \cdots & 0 \\
\cdot & \cdot & \cdot & \cdot & \cdot & \cdot & \cdot & \cdot & \cdot & \cdot & \cdot \\
p^m(\boldsymbol{x}_1) & \cdots & p^m(\boldsymbol{x}_i) & \cdots & p^m(\boldsymbol{x}_n) & 0 & 0 & \cdots & 0 & \cdots & 0
\end{bmatrix}$$

and $\gamma(\cdot)$ is the **semivariogram** of the residuals. The **estimation variance**, $\sigma_E^2(\boldsymbol{x}_0)$, is given by

$$\sigma_E^2(\boldsymbol{x}_0) = \boldsymbol{X}^T\boldsymbol{B}$$

In contrast to **residual kriging**, universal kriging circumvents the need to estimate the drift coefficients a_j. The universal kriging system of equations is identical to that of **intrinsic kriging** using the same drift **model**. The difference between the two is in the determination of the **measure** of spatial continuity.

UNIVERSE. In statistics, an alternative term for **population**.

UPPER QUARTILE. The 0.75 **quantile**.

VALIDATION. A procedure that provides, by reference to independent sources, evidence that a study conforms to its declared purpose. Discrepancies are an indication of deficiencies that require corrective action. Unfortunately, the absence of discrepancies is inconclusive as offsetting **errors** may cancel out and provide a reasonable validation. In either circumstance, a validation does not isolate a specific source of trouble, which must be identified heuristically through trial and error. In general, because of the lack of alternatives, a good validation exercise is interpreted as a sign of sound work. Most validation methods used in **geostatistics** are variations of **cross-validation**.

VARIANCE. For a **random variable** X, the variance $\mathrm{Var}(X)$ is the **expected value** $\mathrm{E}[(X-\mu)^2]$, where μ is the **mean**. The variance is a **measure** of dispersion around the mean. If $(x_1, x_2, \ldots, x_i, \ldots, x_n)$ are values in a **sample** of n observations, the sample variance is the **statistic**

$$s^2 = \frac{1}{n} \sum_{i=1}^{n} (x_i - m)^2 = \frac{1}{n} \left[\sum_{i=1}^{n} x_i{}^2 \right] - m^2$$

where m is the **arithmetic mean** of the observations. The value $\frac{n}{n-1} s^2$ is an **unbiased** type of **estimate** of the **population** variance. The sample variance usually is not **robust** to **outliers**.

VARIANCE ADJUSTMENT FACTOR. Let v_1 and v_2 be two different **supports** for the same attribute. If v_1 is smaller than v_2, then the variance adjustment factor is the reduction in the **variance** of the **distribution** associated with the larger support relative to the variance of the distribution with the smaller support. This factor, f, is related to the **mean semivariogram** $\bar{\gamma}(\cdot,\cdot)$ through

$$f = 1 - \frac{\bar{\gamma}(v_1, v_1)}{\bar{\gamma}(v_2, v_2)}$$

VARIATE. A synonym for **random variable**.

VARIOGRAM. (a) Rigorously, twice the **semivariogram**. (b) For short in the notation, variogram is sometimes used as a synonym for semivariogram.

VARIOGRAM OF ORDER $\boldsymbol{\alpha}$. A **measure** of the degree of dissimilarity in the **outcome** of sampling as the distance between the observation sites increases. Given a **random function** $Z(\cdot)$ that is **intrinsic** or **stationary**, the variogram of order α is the **moment**

$$\mathrm{E}[|Z(\boldsymbol{x}+\boldsymbol{h}) - Z(\boldsymbol{x})|^{\alpha}]$$

where α is a real number larger than 0; and $\boldsymbol{x}$ and $(\boldsymbol{x}+\boldsymbol{h})$ are two locations in the **field**. Any **model** for a variogram of order α satisfies the **permissibility criteria** to be a model for a variogram of order $\beta \le \alpha$. Half the variogram of order 0.5 is called the **rodogram**, the variogram of order 1 is called the **madogram**, and half the variogram of order 2 is the **semivariogram**.

VARIOGRAPHY. The inference of a **model** for a **semivariogram**.

VECTOR. An ordered set of n components $(x_1, x_2, \ldots, x_n)$ subject to the operations of addition and multiplication. The number of components may be finite or infinite. Geometrically, a vector can be regarded as an oriented segment in a **euclidean space** going from the origin to a point whose coordinates are given by the components. A vector is usually denoted by a bold italic lower case letter, such as $\boldsymbol{x}$. This convention, adopted for this glossary, should not be confused with bold, italic capital letters, like $\boldsymbol{X}$, which are commonly used to represent **matrices**. However, vectors also can be viewed as a special case of matrices with either one column or one row. Other conventions used elsewhere to represent vectors include an overbar in the form of an arrow $\vec{x}$ and an underscore $\underline{x}$.

VECTOR SPACE. An alternative term for **linear space.**

VORONOI DIVISION. Let $Y=\{y_1,y_2,...,y_i,...\}$ be an infinite set of points in an n-dimensional **euclidean space** and let $X=\{x_1,x_2,...,x_i,...,x_k\}$ be a subset of Y in which no two points coincide. A Voronoi division is the unique **partition of the set** Y into subsets $v(x_i)$ such that all points in $v(x_i)$ are closer to x_i than to any other point in X. **Poisson processes** are used to generate synthetic sets of X. The complexity of the division increases at a high rate with the dimensionality n of the space. The one-dimensional case is trivial; the region associated with each point extends along the line in each direction from the point halfway to the next point, or to infinity at the end points. Two-dimensional **Voronoi polygons** represent the most widely used case. Three-dimensional polyhedra are so difficult to generate that their applications are limited in spite of interest in their use. A classical three-dimensional example is the growth of crystals with random nucleation. If all crystals start growing simultaneously at the same rate, the Voronoi polyhedra represent the volume available for growth to each nucleus. Voronoi polyhedra are **closed** and convex. The **Delaunay subdivision** is a dual subdivision of the Voronoi division.

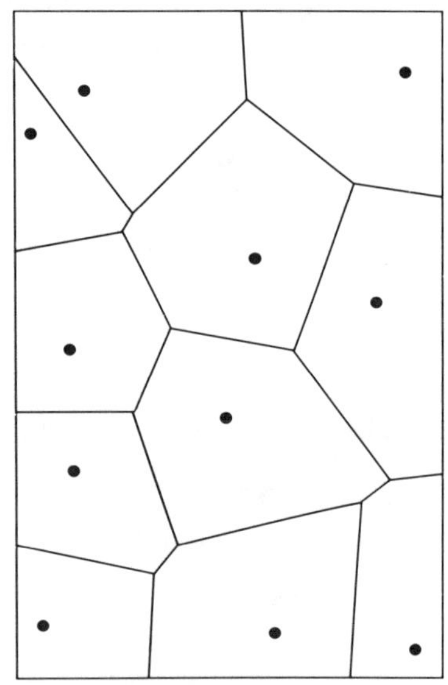

VORONOI POLYGON. Any one of the **partitions** in a two-dimensional **Voronoi division**. Voronoi polygons have been used in the **estimation** of **regionalized variables** $z(x)$. Given a set of point values in the plane $\{x_1,x_2,...,x_i,...,x_k\}$, the value $z(x_i)$ is taken as the **estimate** for all locations in the polygon $v(x_i)$. The procedure has been used widely for estimation in mining in spite of the fact that the method provides satisfactory results only for highly continuous variables and tends to perform poorly in terms of **conditional unbiasedness**. The roots of **geostatistics** are related to efforts to produce better estimates than those provided by the polygonal method.

VORONOI TESSELATION. A two-dimensional **Voronoi division.**

WEAK STATIONARITY. A property such that the **expected value** of a **random process** is a constant with a **covariance** depending only on the **lag** τ. Nothing is implied about **joint probability density functions**. Clearly, a process that has **stationarity** of order 2 is **weakly stationary**; however, the converse is not true.

WHITE NOISE. A **stochastic process** whose **random variables** are **statistically independent**. The **correlogram** of white noise is the spike $\mathrm{Cov}(0)[H(r)-1]^2$, where Cov(0) is the **a priori variance** and $H(r)$ is the **Heaviside's function**.

WHITE RANDOM FUNCTION. The equivalent of **white noise** in n-dimensions.

WIENER FILTER. A minimum **mean square error estimator** of a **stochastic process** $Z(t)$ which is non**recursive**, linear, and **distribution**-free

$$z^*(t_0)=\sum_{i=0}^{n}\lambda_i\, y(t_i)$$

Here $z^*(t_0)$ is the **estimate** of the process $Z(t)$ at time t_0, based on n observations $y(t_i)$ taken at times t_i and that may contain measurement **errors**. The λ_i are unknown weights that can be estimated by solving the system of **normal equations**

$$\begin{bmatrix} \lambda_1 \\ \cdots \\ \lambda_i \\ \cdots \\ \lambda_n \end{bmatrix} = \begin{bmatrix} \rho_y(0) & \cdots & \rho_y(t_1,t_i) & \cdots & \rho_y(t_1,t_n) \\ \cdot & \cdot & \cdot & \cdot & \cdot \\ \rho_y(t_i,t_1) & \cdots & \rho_y(0) & \cdots & \rho_y(t_i,t_n) \\ \cdot & \cdot & \cdot & \cdot & \cdot \\ \rho_y(t_n,t_1) & \cdots & \rho_y(t_n,t_i) & \cdots & \rho_y(0) \end{bmatrix}^{-1} \begin{bmatrix} \rho_{yz}(t_1,t_0) \\ \cdots \\ \rho_{yz}(t_i,t_0) \\ \cdots \\ \rho_{yz}(t_n,t_0) \end{bmatrix}$$

where $\rho_y(\cdot)$ denotes the **correlation coefficient** of the **data** and $\rho_{yz}(\cdot)$ is the correlation coefficient between the estimate and the input data.

WIENER–LEVY PROCESS. An alternative term for **Wiener process.**

WIENER PROCESS. A **random process** $\{Z(t), t \in T\}$ is said to be a Wiener process if

i. The process has **stationary** independent increments,

ii. $Z(t)$ is **normally distributed** for every $t > 0$,

iii. $E[Z(t)] = 0$, for all $t > 0$,

iv. $Z(0) = 0$

The process has an infinite **variance** and thus no **covariance**, yet the **semivariogram** exists and is linear. The Wiener process was originally formulated to **model** the **Brownian motion.**

WIGNER–SEITZ CELL. A synonym for **Voronoi polygon.**

WINDOW. A portion of a **field.**

WORKING NEIGHBORHOOD. A **neighborhood** for a volume under study. Changes in size and shape of the working neighborhood should be minimized to assure consistency in a study. The size of a working neighborhood determines the **working scale**. The **kriging neighborhood** is the most frequently used working neighborhood.

WORKING SCALE. The **average** size of a **working neighborhood**. The working scale determines the validity of a **model** as well as inherent characteristics such as **stationarity**. For a given **regionalized variable**, the order of the associated **intrinsic random function** will vary with the choice of the working scale in a way that is not easily **predicted.**

ZONAL ANISOTROPY. (a) A special case of **nested structures** that results in at least one **structure** that is not a **function** of all n dimensions of the **euclidean space** in which the **regionalized variable** has been defined. In three dimensions, a simple example of a **nested semivariogram** $\gamma(\boldsymbol{h})$ that displays zonal anisotropy is

$$\gamma(\boldsymbol{h}) = \gamma_1(\boldsymbol{h}) + \gamma_2(h_3)$$

(b) A nested structure in which each component structure may have its own form of **anisotropy.**

ZONE OF INFLUENCE. For a **random variable** $Z(\boldsymbol{x}_0)$, that portion of the **field** around location $\boldsymbol{x}_0$ whose random variables $Z(\boldsymbol{x}_0 + \boldsymbol{h})$ are **statistically dependent** on $Z(\boldsymbol{x}_0)$. If only linear **stochastic dependence** is considered, the radius of the zone of influence is equal to the **range.**

Z-SCORE. A value of a rescaled **variate** that is obtained by subtracting the sample **mean** and dividing the difference by the **standard deviation** of the **sample**. The z-score is also called the *standard score*. A **reduced residual** can be viewed as a form of z-score.

SELECTED BIBLIOGRAPHY

The preparation of this glossary required reading a large number of books, papers, dissertations, thesis, notes, and reports, as well as attending many technical presentations. Almost every term included here was used more than once by diverse authors in different places. The sources listed include only those works that were eventually used to develop the definition or the translation of at least one term. Thus, the bibliography is neither a detailed inventory of all the material that was actually surveyed nor a comprehensive account of all that has been published under the heading of geostatistics, whatever may be the scope of geostatistics. The selection is deliberately biased toward books because they are the published references that cover the most ground of any form of publication; this minimizes the number of entries in the list. Books also are the most thoroughly reviewed of all written publications, which helps to reduce the number of errors and oddities. Finally, books tend to be more widely available, in case the reader decides to go beyond this glossary for more information. The list is not restricted to geostatistical works. Although geostatistical sources were used both to select the terms in the glossary and, in many instances, to provide the definitions, additional nongeostatistical publications were also used in preparing the definitions. These sources were referred to for the more basic terms that did not originate within geostatistics, but rather in more traditional disciplines such as functional analysis, calculus, and statistics.

AGTERBERG, F.P. 1974. *Geomathematics. Developments in Geomathematics* 1. Elsevier Scientific Publishing Company, Amsterdam, 596 pp.

AKIN, H. and H. SIEMES. 1988. *Praktische Geostatistik—Eine Einführung für den Bergbau und die Geowissenschaften.* Springer-Verlag, Berlin, 304 pp.

ALFARO, M. 1980. The random coin method. Solution of the problem of the simulation of a random function in the plane. *Mathematical Geology* 12(1):25–32.

ALFARO, M. and F. MIGUEZ. 1976. Optimal interpolation using transitive methods. In *Advanced Geostatistics in the Mining Industry*, M. GUARASCIO, M. DAVID, and C. HUIJBREGTS, eds. Proceedings of the NATO Advanced Study Institute held at the University of Rome, Italy, October 13–25, 1975. D. Reidel Publishing Company, Dordrecht, The Netherlands, pp. 91–99.

ARMSTRONG, M. and G. MATHERON. 1986. Disjunctive kriging revisited. *Mathematical Geology* 18(8):711–742.

BERBERIAN, S.K. 1961. *Introduction to Hilbert Space.* Oxford University Press, New York, 206 pp.

CARR, J.R. and D.E. MYERS. 1985. COSIM: a FORTRAN IV program for coconditional simulation. *Computers & Geosciences* 11(6):675–705.

CHAUTRU, J.M. 1989. The use of Boolean random functions in Geostatistics. In *Geostatistics*, M. ARMSTRONG, ed. Proceedings of the Third International Geostatistics Congress, Avignon, France, September 5–9, 1988. Kluwer Academic Publishers, Vol. 1, pp. 201–212.

CHERNUKHIN, A.E. 1962. *English-Russian Technical Dictionary.* Central Editorial Board, Foreign-Language Scientific and Technical Dictionaries FIZMATGIZ, Moscow, 635 pp.

CHRISTAKOS, G. 1984. On the problem of permissible covariance and variogram models. *Water Resources Research* 20(2):251–265.

CHRISTAKOS, G. 1988. Stochastic simulation of spatially correlated geoprocesses. *Mathematical Geology* 19(8):807–831.

CHRISTAKOS, G. 1989. Optimal estimation of nonlinear state nonlinear observation systems. *Journal of Optimization Theory and Applications* 62(1):29–48.

CLARK, I. 1979. *Practical Geostatistics.* Applied Science Publishers Ltd., London, 129 pp.

CONSIDINE, D.M., ed. 1983. *Van Nostrand Scientific Encyclopedia*, sixth edition. Van Nostrand-Reinhold Company, New York, 3067 pp.

CRESSIE, N. 1986. Kriging nonstationary data. *Journal of the American Statistical Association* 81(395):625–634.

CRESSIE, N. 1987. A nonparametric view of generalized covariances for kriging. *Mathematical Geology* 19(5):425–449.

DAVIS, M.W. 1987. Production of conditional simulations via the LU triangular decomposition of the covariance matrix. *Mathematical Geology* 19(2):91–98.

DAVID, M. 1977. *Geostatistical Ore Reserve Estimation. Developments in Geomathematics* 2. Elsevier Scientific Publishing Company, Amsterdam, 364 pp.

DAVID, M. 1978. Sampling and estimation problems for three dimensional spatial stationary and nonstationary stochastic processes as encountered in the mineral industry. *Journal of Statistical Planning and Inference* 2:211–244.

DAVID, M. 1988. *Handbook of Applied Advanced Geostatistical Ore Reserve Estimation. Developments in Geomathematics* 6. Elsevier Science Publishers B.V., Amsterdam, 216 pp.

DEMANGE, C., C. LAJAUNIE, C. LANTUEJOUL, and J. RIVOIRARD. 1987. Global recoverable reserves. Testing various changes of support models on uranium data. In *Geostatistical Case Studies*, G. MATHERON and M. ARMSTRONG, eds. D. Reidel Publishing Company, Dordrecht, The Netherlands, pp. 187–208.

EISENREICH, G. 1982. R. Sube s avtorskym kolektivom - MATEMATIKA Anglicko-Nemecko-Francúzsko-Rusko-Slovenskij Slovnik (A–Z) ALFA vydavatelstvo technickej a ekonomickej literatury. Bratislava, VEB Verlag Technik, Berlin, 923 pp.

ENGLUND, E. and A. SPARKS. 1988. *Geostatistical Environmental Assessment Software. User's Guide*. U.S. Environmental Protection Agency, Las Vegas, 172 pp.

FREIBERGER, W.F., ed. 1960. *The International Dictionary of Applied Mathematics*. D. Van Nostrand Company, Inc., Princeton, N.J., 1173 pp.

FROIDEVAUX, R. 1988. *Geostat Toolbox Primer. Version 1.10*. Froidevaux, Srivastava, Schofield & Associates (FSS), Vancouver, 88 pp.

GALLI, A. and G. MEUNIER. 1987. Study of a gas reservoir using the external drift method. In *Geostatistical Case Studies*, G. MATHERON and M. ARMSTRONG, eds. D. Reidel Publishing Company, Dordrecht, The Netherlands, pp. 105–119.

GALLI, A. and L. SANDJIVY. 1985. Analyse krigeante et analyse spectrale. *Sciences de la Terre, Série Informatique Géologique* 21:115–124.

GANDIN, L.S. 1963. *Ob"ektivnyi Analiz Meteorologicheskikh Polei. Gidrometeorologicheskoe Izdatel'stvo (GIMIZ)*, Leningrad. Translated into English as *Objective Analysis of Metereological Fields* in 1965 by the Israel Program for Scientific Translations, Jerusalem, Israel, and published for The U.S. Department of Commerce and The National Science Foundation, Washington, D.C., 242 pp.

GELLERT, W., H. KÜSTNER, M. HELLWICH, and H. KÄSTNER, eds. 1977. *The VNR Concise Encyclopedia of Mathematics*. Van Nostrand-Reinhold Company, New York, 816 pp.

GOLDBERGER, A.S. 1962. Best linear unbiased prediction in the generalized linear regression model. *Journal of the American Statistical Association* 57(298):369–375.

GOULD, P. 1967. On the geographic interpretation of eigenvalues. An initial exploration. *Transactions of the Institute of British Geographers* 42:53–86.

GROESTSCH, C.W. 1980. *Elements of Applicable Functional Analysis*. Marcel Dekker, Inc., New York, 300 pp.

HOGG, R.V. and A.T. CRAIG. 1970. *Introduction to Mathematical Statistics*. The Macmillan Company, New York, 415 pp.

HOHN, M.E. 1988. *Geostatistics and Petroleum Geology*. Van Nostrand-Reinhold Company, New York, 264 pp.

ISAAKS, E.H. and R.M. SRIVASTAVA. 1988. Spatial continuity measures for probabilistic and deterministic geostatistics. *Mathematical Geology* 20(4):313–341.

ISAAKS, E.H. and R.M. SRIVASTAVA. 1989. *An Introduction to Applied Geostatistics*. Oxford University Press, New York, 561 pp.

ITO, K., ed. 1987. *Encyclopedic Dictionary of Mathematics*. English translation of the third Japanese edition by the Massachusetts Institute of Technology Press, Cambridge, Massachusetts, 2148 pp. (4 volumes). Originally published in Japanese in 1954 under the title *Iwanami Sugaku Ziten* by Iwanami Shoten Publishers, Tokyo.

JAMES, G. and R.C. JAMES. 1976. *Mathematics Dictionary*. Van Nostrand-Reinhold Company, New York, 509 pp.

JOURNEL, A.G. 1984. The place of non-parametric geostatistics. In *Geostatistics for Natural Resources Characterization*, G. VERLY, M. DAVID, A.G. JOURNEL, and A. MARÉCHAL, eds. Proceedings of the Second NATO Advanced Study Institute held in South Lake Tahoe, California, September 1983. D. Reidel Publishing Company, Dordrecht, The Netherlands, Part 1, pp. 307–335.

JOURNEL, A.G. 1985. The deterministic side of geostatistics. *Mathematical Geology* 17(1):1–15. Discussion in 17(8):855–869.

JOURNEL, A.G. 1988. New distance measures. The route toward truly non-Gaussian Geostatistics. *Mathematical Geology* 20(4):459–475.

JOURNEL, A.G. 1989. *Fundamentals of Geostatistics in Five Lessons.* Short course presented at the 28th International Geological Congress, Washington, D.C. *Short Course In Geology*, Vol. 8, American Geophysical Union, Washington, D.C., 40 pp.

JOURNEL, A.G. and F.G. ALABERT. 1988. Focusing on spatial connectivity of extreme-valued attributes. Stochastic indicator models of reservoir heterogeneities. *Paper SPE 18324* prepared for presentation at the 63rd Annual Technical Conference and Exhibition of the Society of Petroleum Engineers, Society of Petroleum Engineers, Richardson, Texas, 12 pp.

JOURNEL, A.G. and C.J. HUIJBREGTS. 1978. *Mining Geostatistics*. Academic Press, London, 600 pp.

KENDALL, M.G. and W.R. BUCKLAND. 1971. *A Dictionary of Statistical Terms*, third edition. Oliver and Boyd, Edinburgh, 166 pp.

KIM, Y.C., D.E. MYERS, and H.P. KNUDSEN. 1977. *Advanced Geostatistics in Ore Reserve Estimation and Mine Planning (Practitioner's Guide)*. Report to the United States Department of Energy, Grand Junction, Colorado, 236 pp.

KOTZ, S., N.L. JOHNSON, and C.B. READ, eds. 1988. *Encyclopedia of Statistical Sciences*. A Wiley-Interscience Publication, New York, 6317 pp. (9 volumes).

KRIGE, D.G. 1976. A review of the developments of geostatistics in South Africa. In *Advanced Geostatistics in the Mining Industry*, M. GUARASCIO, M. DAVID, and C. HUIJBREGTS, eds. Proceedings of the NATO Advanced Study Institute held at the University of Rome, Italy, October 13–25, 1975. D. Reidel Publishing Company, Dordrecht, The Netherlands, pp. 279–311.

LEMMER, I.C. 1986. Mononodal indicator variography. *Mathematical Geology* 18(7):589–623.

LUENBERGER, D. 1969. *Optimization by Vector Space Methods*. John Wiley & Sons, Inc., New York, 326 pp.

LUSTER, G.R. 1986. Homogenization and proportioning of mined materials. Application of conditional simulation of coregionalization. In *Proceedings of the 19th International Symposium on the Application of Computers and Operations Research in the Mineral Industry (APCOM)*, R.V. Ramani, ed. Society of Mining Engineers, Inc., Littleton, Colorado, pp. 163–172.

MACLANE, S. and G. BIRKHOFF. 1970. *Algebra*, fourth edition. The Macmillan Company, New York, 598 pp.

MANDELBROT, B.B. 1983. *The Fractal Geometry of Nature*. Updated and augmented. W.H. Freeman and Company, New York, 468 pp.

MARÉCHAL, A., ed. 1980. *Geostatistical Ore Reserve Estimation Library. User's Guide*. Centre de Géostatistique et de Morphologie Mathématique, Fontainebleau, France, 95 pp.

MARSILY, G. DE. 1986. *Quantitative Hydrogeology*. Academic Press, Inc., Orlando, 440 pp.

MARTIN, M., M. BORATYN, J. JANUSZKIEWICZ, A. MURAWSKA, J. SZARSKI, and E. TABACZKIEWICZ. 1980. *Wielki Slownik Techniczny Rosyjsko-Polski*. Wydawnictwa Naukowo-Techniczne, Wrocław, Poland, 1151 pp.

MATÉRN, B. 1960. Spatial variation. *Meddelanden från Statens Skogsforskningsinstitut* 49(5):1–144. A second edition is also available since 1986 as *Lecture Notes in Statistics*, no. 36, Springer-Verlag, Berlin, 151 pp.

MATHERON, G. 1962. Traité de Géostatistique Appliquée. *Mémoires du Bureau de Recherches Géologiques et Minières*, Vol. 14. Editions Technip, Paris, 333 pp.

MATHERON, G. 1965. *Les variables régionaliseés et leur estimation*. Masson et Cie., Editeurs, Paris, 305 pp.

MATHERON, G. 1968. *Osnovy Prikladnoï Geostatistiki*. Mir, Moscow, 408 pp.

MATHERON, G. 1970. *La théorie des variables régionalisées, et ses applications*. Les Cahiers du Centre de Morphologie Mathématique de Fontainebleau, Fascicule 5, Ecole Nationale Supérieure des Mines, Paris, 211 pp. English version available from the same publisher since 1971.

MATHERON, G. 1973. The intrinsic random functions and their applications. *Advances in Applied Probability* 5:439–468.

MATHERON, G. 1975. *Random Sets and Integral Geometry*. John Wiley and Sons, New York, 260 pp.

MATHERON, G. 1978. *Estimer et choisir*. Les Cahiers du Centre de Morphologie Mathématique de Fontainebleau, Fascicule 7, Ecole Nationale Supérieure des Mines, Paris, 175 pp. English translation by A.M. HASOFER available since 1989, *Estimating and Choosing. An Essay on Probability in Practice*. Springer-Verlag, Berlin, 141 pp.

MATHERON, G. 1981. Splines and kriging: their formal equivalence. *Syracuse University Geology Contribution*, Vol. 8, pp. 77–95.

MORAN, P.A.P. 1968. *An Introduction to Probability Theory*. Oxford University Press, London, 542 pp.

MÜLLER, V.K. 1977. *English-Russian Dictionary*. Russian Language Publishers, Moscow, 887 pp.

MÜLLER, P.H. 1975. *Wahrscheinlichkeitsrechnung und Mathematische Statistik. Lexikon der Stochastik*. Akademie-Verlag, Berlin, 443 pp.

MYERS, D.E. 1982. Matrix formulation of co-kriging. *Mathematical Geology* 14(3):249–257.

MYERS, D.E. 1989. To be or not to be...stationary? That is the question. *Mathematical Geology* 21(3):347–362.

NEUMAN, S.P. and E.A. JACOBSON. 1984. Analysis of nonintrinsic spatial variability by residual kriging with application to regional groundwater levels. *Mathematical Geology* 16(5):499–521.

ORLOV, V.B., N.S. SKOROKHOD, and A.B. SOSINSKY. 1987. *Russian-English-German-French Mathematical Dictionary*. Russky YAZYK Publishers, Moscow, 299 pp.

PAPOULIS, A. 1965. *Probability, Random Variables, and Stochastic Processes*. McGraw-Hill Book Company, New York, 583 pp.

PARKER, S.P., ed. 1984. *Dictionary of Scientific and Technological Terms*, third edition. McGraw-Hill Book Company, New York, 1849 pp.

PARZEN, E. 1965. *Stochastic Processes*, third edition. Holden-Day Inc., San Francisco, 324 pp.

PEARSON, C.E., ed. 1974. *Handbook of Applied Mathematics*. Van Nostrand-Reinhold Company, New York, 1265 pp.

RENDU, J.-M. 1978. *An Introduction to Geostatistical Methods of Mineral Evaluation*. South African Institute of Mining and Metallurgy, Johannesburg, 84 pp.

ROYLE, A.G., I. CLARK, P.I. BROOKER, H. PARKER, A.G. JOURNEL, J.M. RENDU, R. SANDEFUR, D.C. GRANT, and P. MOUSSET-JONES. 1980. *Geostatistics*. McGraw-Hill Book Company, New York, 168 pp.

SACHS, L. 1971. *Statistische Auswertungsmethoden*. Springer-Verlag, Berlin, 545 pp.

SHERIFF, R.E. 1984. *Encyclopedic Dictionary of Exploration Geophysics*. Society of Exploration Geophysics, Tulsa, Oklahoma, 323 pp.

VERLY, G. 1984. The block distribution given a point multivariate normal distribution. In *Geostatistics for Natural Resources Characterization*, G. VERLY, M. DAVID, A.G. JOURNEL, and A. MARÉCHAL, eds. Proceedings of the Second NATO Advanced Study Institute held in South Lake Tahoe, California, September 1983. D. Reidel Publishing Company, Dordrecht, The Netherlands, Part 1, pp. 495–515.

WEBSTER, R. and M.A. OLIVER. 1989. Disjunctive kriging in agriculture. In *Geostatistics*, M. ARMSTRONG, ed. Proceedings of the Third International Geostatistics Congress, Avignon, France, September 5–9, 1988. Kluwer Academic Publishers, Vol. 1, pp. 421–432.

WEIK, M.H. 1970. *Standard Dictionary of Computers and Information Processing*. Hayden Book Company, New York, 326 pp.

WHITTLE, P. 1963. *Prediction and Regulation by Linear Least-Square Methods*. English University Press Ltd., London, 143 pp. Reprinted by the University of Minnesota Press, Minneapolis.

WIENER, N. 1949. *The Extrapolation, Interpolation, and Smoothing of Stationary Time Series with Engineering Applications*. John Wiley & Sons, Inc., New York, 163 pp.

YAGLOM, A.M. 1962. *An Introduction to the Theory of Stationary Random Functions*. Prentice-Hall, Inc., Englewood Cliffs, N.J., 235 pp.

PART II

THE MULTILINGUAL DICTIONARY

English	Chinese	French	German
absolute dispersion	绝对离散	dispersion absolue	absolute Dispersion
accumulation	累积	accumulation	Akkumulation
accumulation point	聚点	point d'accumulation	Häufungspunkt
accuracy	准确度	exactitude	Genauigkeit
additive set function	可加集函数	fonction d'ensemble additive	additive Mengenfunktion
additivity theorem	可加性定理	théorème d'additivité	Additionssatz, -theorem
affine correction	仿射校正	correction affine	affine Korrektur
affinity transformation	仿射变换	affinité	affine Transformation
aleatory variable	随机变量	variable aléatoire	Zufallsvariable
anamorphosis	变形	anamorphose	Anamorphismus, Anamorphose
anisotropy	各向异性	anisotropie	Anisotropie
apparent discontinuity at the origin	原点处的表观不连续性	discontinuité apparente à l'origine	scheinbare Diskontinuität im Ursprung
a priori variance	先验方差	variance a priori	a-priori-Varianz
arithmetic average	算术平均值	moyenne arithmétique	arithmetisches Mittel
arithmetic mean	算术平均值	moyenne arithmétique	arithmetisches Mittel
authorized coefficients of order ν	ν阶允许系数	coefficients autorisés d'ordre ν	zulässige Koeffizienten der Ordnung ν
authorized combinations of order ν	ν阶允许组合	combinaisons autorisées d'ordre ν	zulässige Kombinationen der Ordnung ν
autocorrelation	自相关	autocorrelation	Autokorrelation
autocorrelation coefficient	自相关系数	coefficient d'autocorrélation	Autokorrelationskoeffizient
autocorrelation function	自相关函数	fonction d'autocorrélation	Autokorrelationsfunktion
autocorrelogram	自相关图	autocorrélogramme	Autokorrelogram
autocovariance	自协方差	autocovariance	Autokovarianz
autokrigeability	自克里格性	autokrigeabilité	Autokrigbarkeit
autokriging	自克里格法	autokrigeage	Autokrigen
average	平均	moyenne	Mittel, Mittelwert
average semivariogram	平均半变差函数(图)	variogramme moyen	mittleres Semivariogramm
average value	平均值	moyenne	Mittelwert
basis function	基函数	fonction de base	Basisfunktion
Bayes' theorem	贝叶斯定理	théorème de Bayes	Satz von Bayes
Bernoulli experiment	贝努里试验	tirages de Bernoulli	Bernoulli-Versuch
Bessel inequality	贝塞尔不等式	inégalité de Bessel	Besselsche Ungleichung
beta distribution	β分布	loi Beta	Betaverteilung
beta function	β函数	fonction Beta	Betafunktion
bias	偏倚	biais	Verzerrung
bigaussian model	二元高斯模型	loi bigaussienne	zweidimensionales Gauß-sches Modell
bijection	双射	bijection	bijektive Abbildung, eineindeutige Abbildung

GREEK, PORTUGUESE, RUSSIAN, and SPANISH

Greek	Portuguese	Russian	Spanish
απόλυτη διασπορά	dispersão absoluta	абсолютная дисперсия	dispersión absoluta
συσσώρευση	acumulação	аккумуляция, накопление	acumulación
σημείο συσσώρευσης	ponto de acumulação	точка накопления, предельная точка	punto de acumulación
ακρίβεια	exactidão	точность	exactitud
προσθετική συνολο-συνάρτηση	função de conjunto aditiva	функция аддитивного множества	función de conjunto aditiva
προσθετικό θεώρημα	teorema da aditividade	аддитивная теорема	teorema de la aditividad
συσχετισμένη διόρθωση	correcção afim	аффинная коррекция	corrección afín
συσχετισμένος μετασχηματισμός	transformação afim	аффинное преобразование	transformación afín
τυχαία μεταβλητή	variável aleatória	случайная величина, переменная	variable aleatoria
αναμόρφωση	anamorfose	анаморфоз	anamorfosis
ανισοτροπία	anisotropia	анизотропия	anisotropía
φαινομενική ασυνέχεια στήν αρχή	descontinuidade aparente na origem	видимая разрывность в начале координат	discontinuidad aparente en el origen
προτέρα διασπορά	variância à priori	априорная дисперсия	varianza poblacional
αριθμητική μέση τιμή	média aritmética	среднее арифметическое	promedio aritmético
αριθμητικός μέσος όρος	média aritmética	средняя арифметическая величина	media aritmética
επιτρεπτοί συντελεστές τάξης v	coeficientes autorizados de ordem v	допустимые коэффициенты порядка v	coeficientes autorizados de orden v
επιτρεπτοί συνδιασμοί τάξης v	combinações autorizadas de ordem v	допустимые комбинации порядка v	combinaciones autorizadas de orden v
αυτοσυσχέτιση	autocorrelação	автокорреляция	autocorrelación
συντελεστής αυτοσυσχέτισης	coeficiente de autocorre-lação	коэффициент автокорре-ляции	coeficiente de autoco-rrelación
συνάρτηση αυτοσυσχέτισης	função de autocorrelação	автокорреляционная функция	función de autocorrelación
αυτοσυσχετόγραμμα	autocorrelograma	автокоррелограмма	autocorrelograma
αυτοσυνδιασπορά	autocovariância	автоковариация	autocovarianza
αυτοκρίκιγκ	autokrigabilidade	способность к авто-крайгингу	autokrigeabilidad
αυτοκρίκιγκ	autokrigagem	автокрайгинг	autokrigeage
μέσος όρος	média	среднее	promedio
μέσο ημιβαριόγραμμα	variograma médio	средняя семивариограмма	semivariograma promedio
μέση τιμή	valor médio	среднее значение, средняя величина	valor promedio
συνάρτηση βάσης	função de base	базисная функция	función de base
θεώρημα Bayes	teorema de Bayes	теорема Байеса	teorema de Bayes
πείραμα Bernoulli	provas de Bernoulli	эксперимент Берноулли	experimento de Bernoulli
ανισότης Bessel	desigualdade de Bessel	неравенство Бесселя	desigualdad de Bessel
κατανομή Beta	distribuição Beta	бета-распределение	distribución beta
συνάρτηση Beta	função Beta	бета-функция	función beta
μεροληψία	enviesamento	смещение	sesgo
δικανονική κατανομή	modelo bigaussiano	бигауссовская модель	modelo bigaussiano
αμφιμονότιμη συνάρτηση	bijecção	биекция, изоморфизм	biyección

English	Chinese	French	German
bimodal distribution	双峰分布	distribution bimodale	bimodale Verteilung
binomial distribution	二项分布	loi binomiale	Binomialverteilung
bivariate	二元的	bivariable	zweidimensional
bivariate distribution	二元分布	loi bivariable	zweidimensionale Verteilung
bivariate normal distribution	二元正态分布	loi normale bivariable	zweidimensionale Normalverteilung
block kriging	块段克里格法	krigeage de blocs	Block-Krigen
Boolean random function	布尔随机函数	schéma booléen, fonction aléatoire booléenne	Boolesche Zufallsfunktion
Borel field	波莱尔域	champ Borelien	Borel-Feld
Brownian motion process	布朗运动过程	mouvement Brownien	Brownsche Bewegung
Buniakovski inequality	布尼亚科夫斯基不等式	inegalité de Buniakovski	Buniakovskische Ungleichung
Cartier's relation	卡蒂尔关系式	relation de Cartier	Cartiersche Formel
Cauchy–Schwarz inequality	柯西-许瓦尔兹不等式	inegalité de Cauchy-Schwarz	Cauchy-Schwarzsche Ungleichung
Cauchy sequence	柯西序列	suite de Cauchy	Cauchy-Folge
centered covariance	中心协方差	covariance centrée	zentrierte Kovarianz
centile	百分位数	centile	Zentil
central limit theorem	中心极限定理	théorème de la limite centrale	zentraler Grenzwertsatz
change of support model	承载变化模型	modèle de changement de support	Modell der Stützungsänderung
characteristic equation of a matrix	矩阵的特征方程	équation caractéristique d'une matrice	charakteristische Gleichung einer Matrix
Choleski's method	乔勒斯基方法	méthode de Choleski	Choleskische Methode
Choquet's capacity	乔奎特容量	capacité de Choquet	Choquetsche Eigenschaft
closed	闭合的	fermé	abgeschlossen
coconditional simulation	协条件模拟	cosimulation conditionnelle	bedingte Kosimulation
codimension	余维数	codimension	Kodimension
coefficient of excess	超出系数	coefficient d'excès	Exzeß
coefficient of kurtosis	峰态系数	coefficient d'aplatissement	Wölbung
coefficient of skewness	偏态系数	coefficient d'asymétrie	Schiefe
coefficient of variation	变异系数	coefficient de variation	Variationskoeffizient
cofactor of an element in a determinant	行列式元素的余子式	cofacteur d'un élément d'un déterminant	Kofaktor eines Elements in einer Determinante
cokriging	协克里格法	cokrigeage	Kokrigen
column of a determinant	行列式的列	colonne d'un déterminant	Spalte in einer Determinante
column of a matrix	矩阵的列	colonne d'une matrice	Spalte einer Matrix
compact	紧的	compact	kompakt
complete set of orthogonal functions	正交函数的完全集	système complet de fonctions orthogonales	vollständiger Satz orthogonaler Funktionen
complete set of orthogonal polynomials	正交多项式的完全集	système complet de polynômes orthogonaux	vollständiger Satz orthogonaler Polynome

Greek	Portuguese	Russian	Spanish
δίκορφη κατανομή	distribuição bimodal	бимодальное распределение	distribución bimodal
διωνυμική κατανομή	distribuição binomial	биномиальное распределение	distribución binomial
διμεταβλητή	bidimensional	двухмерная переменная	bivariable
διμεταβλητή κατανομή	distribuição bidimensional	двухмерное распределение	distribución bivariable
διμεταβλητή κανονική κατανομή	distribuição normal bidimensional	двухмерное нормальное распределение	distribución normal bivariable
μπλοκ κρίκιγκ	krigagem de blocos	блок-крайгинг	krigeage por bloques
τυχαία συνάρτηση Boole	função aleatória booleana, esquema booleano	Булевская случайная функция	función aleatoria de Bool
πεδίο Borel	campo de Borel	борелево поле	campo de Borel
διαδικασία Brown	movimento Browniano	Броуновский процесс движения	movimiento browniano
ανισότης Buniakovski	desigualdade de Buniakovski	неравенство Буняковского	desigualdad de Buniakovski
σχέση Cartier	relação de Cartier	соотношение Картье	relación de Cartier
ανισότης Cauchy–Schwarz	desigualdade de Cauchy-Schwarz	неравенство Коши-Шварца	desigualdad de Cauchy–Schwarz
ακολουθία Cauchy	sucessão de Cauchy	последовательность Коши	secuencia de Cauchy
κεντρική συνδιασπορά	covariância centrada	центрированная ковариация	covarianza centrada
εκατοστημόριο	percentil	процентиль	percentil
κεντρικό οριακό θεώρημα	teorema do limite central	центральная граничная теорема	teorema del límite central
μοντέλο αλλαγής φορέα	modelo de mudança de suporte	модель изменения суппорта	modelo de cambio de soporte
χαρακτηριστική εξίσωση ενός πίνακα	equação característica de uma matriz	характеристическое уравнение матрицы	ecuación característica de una matriz
μέθοδος Choleski	método de Choleski	метод Холеского	método de Choleski
χωρητικότης Choquet	capacidade de Choquet	ёмкость Шоке	capacitancia de Choquet
κλειστό	fechado	замкнутый	cerrado
δεσμευμένη προσομοίωση	cossimulação condicional	совместное моделирование, соответствующее моделирование	cosimulación condicionada
ομοιοδιάστατος	codimensão	гомологическая коразмерность	codimensión
συντελεστής υπέρβασης	coeficiente de excesso	коэффициент эксцесса	coeficiente de exceso
συντελεστής κύρτωσης	coeficiente de achatamento	коэффициент эксцесса	coeficiente de curtosis
συντελεστής λοξότητας	coeficiente de assimetria	коэффициент асимметрии	coeficiente de asimetría
συντελεστής μεταβλητότητας	coeficiente de variação	коэффициент вариации	coeficiente de variación
συντελεστής ενός στοιχείου μιάς ορίζουσας	componente algébrico de um elemento num determinante	кофактор элемента в определителе	cofactor de un elemento de un determinante
κοκρίκιγκ	cokrigagem	кокрайгинг, соответствующий крайгинг	cokrigeage, cokrigeado
στήλη μιάς ορίζουσας	coluna de um determinante	столбец детерминанта	columna de un determinante
στήλη ενός πίνακα	coluna de uma matriz	столбец матрицы	columna de una matriz
συμπαγής	compacto	компактный	compacto
πλήρες σύνολο ορθογωνικών συναρτήσεων	sistema completo de funções ortogonais	полное множество ортогональных функций	conjunto completo de funciones ortogonales
πλήρες σύνολο ορθογωνικών πολυωνύμων	sistema completo de polinómios ortogonais	полное множество ортогональных многочленов	conjunto completo de polinomios ortogonales

English	Chinese	French	German
complete space	完全空间	espace complet	vollständiger Raum
conditional bias	条件偏倚	biais conditionnel	bedingte Verzerrung
conditional cosimulation	条件协模拟	cosimulation conditionnelle	bedingte Kosimulation
conditional distribution	条件分布	loi conditionnelle	bedingte Verteilung
conditional expectation	条件期望	espérance conditionnelle	bedingte Erwartung
conditionally positive definite of order ν	ν 阶条件正定	défini positif conditionnel d'ordre ν	bedingt positiv definit der Ordnung ν
conditional nonbias	条件无偏	non-biais conditionnel	bedingte Unverzerrtheit
conditional probability	条件概率	probabilité conditionnelle	bedingte Wahrscheinlichkeit
conditional probability density function	条件概率密度函数	fonction de densité conditionnelle	bedingte Wahrscheinlichkeitsdichte
conditional simulation	条件模拟	simulation conditionnelle	bedingte Simulation
conditional unbiasedness	条件无偏性	non-biais conditionnel	bedingte Unverzerrtheit
confidence interval	置信区间	intervalle de confiance	Vertrauensintervall
confidence limit	置信限	limite de confiance	Vertrauensgrenze
configuration	构形	configuration	Konfiguration
conjugate complex number	共轭复数	nombre imaginaire conjugué	konjugiert komplexe Zahl
conservation of law hypothesis	分布守恒假设	hypothèse de permanence	Permanenzhypothese, Hypothese der Erhaltung
consistency	一致性	cohérence	Konsistenz
constraint effect	约束效应	effet de contrainte	Effekt der Rahmenbedingungen
constraint interval	约束区间	intervalle de contrainte	begrenzendes Intervall
continuous random variable	连续型随机变量	variable aléatoire continue	stetige Zufallsvariable
contour	等值线	ligne de contour	Kontour
conventional parameter	约定参数	paramètre conventionnel	vereinbarungsgemäßer Parameter
convex function	凸函数	fonction convexe	konvexe Funktion
convolution	卷积	convolution	Faltung
coregionalization	协区域化	corégionalisation	Koregionalisierung
correlation	相关	corrélation	Korrelation
correlation coefficient	相关系数	coefficient de corrélation	Korrelationskoeffizient
correlogram	相关图	corrélogramme	Korrelogramm
cosimulation	协模拟	cosimulation	Kosimulation
covariance	协方差	covariance	Kovarianz
covariance matrix	协方差矩阵	matrice des covariances	Kovarianzmatrix
cross-correlation	互相关	corrélation croisée	Kreuzkorrelation
cross-correlation coefficient	互相关系数	coefficient de correlation croisée	Kreuzkorrelationskoeffizient
crosscorrelogram	互相关图	corrélogramme croisé	Kreuzkorrelogramm

Greek	Portuguese	Russian	Spanish
πλήρης χώρος	espaço completo	полное пространство	espacio completo
δεσμευμένη μεροληψία	enviesamento condicional	условное смещение	sesgo condicionado
δεσμευμένη συνπροσομοίωση	cossimulação condicional	совместное моделирование, соответствующее моделирование	cosimulación condicionada
δεσμευμένη κατανομή	distribuição condicionada	условное распределение	distribución condicionada
δεσμευμένη μέση τιμή	esperança matemática condicionada	условное ожидание	valor esperado condicionado
δεσμευμένος θετικά ορισμένος ν τάξης	condicionalmente definida positiva de ordem ν	условно положительная определенная порядка ν	condicionalmente definido positivo de orden ν
δεσμευμένη μη-μεροληψία	não enviesamento condicional	условное не-смещение	sesgo nulo condicionado
δεσμευμένη πιθανότης	probabilidade condicionada	условная вероятность	probabilidad condicionada
δεσμευμένη συνάρτηση πυκνότητος πιθανότητος	função densidade de probabilidade condicionada	функция плотности условной вероятности	función de densidad de probabilidad condicionada
δεσμευμένη προσομοίωση	simulação condicional	условное моделирование	simulación condicionada
δεσμευμένη αμεροληψία	não enviesamento condicional	условная несмещенность	sesgo nulo condicionado
διάστημα εμπιστοσύνης	intervalo de confiança	доверительный интервал	intervalo de confianza
όριο εμπιστοσύνης	limite de confiança	доверительный предел	límite de confianza
σχηματισμός	configuração	конфигурация	configuración
συζυγής μιγαδικός αριθμός	número complexo conjugado	сопряженное комплексное число	número complejo conjugado
υπόθεση διατήρησης του νόμου	hipótese de permanência	сохранение гипотезы закона	hipótesis de conservación de la ley
συνέπεια	coerência	консистенция	consistencia
επίδραση περιορισμού	efeito de constrangimento geométrico	ограничение	efecto restricción
διάστημα περιορισμού	intervalo de constrangimentos	интервал условия	intervalo de la restricción
συνεχής τυχαία μεταβλητή	variável aleatória contínua	непрерывная случайная величина	variable aleatoria continua
ισοϋψής	contorno	контур	curva de nivel
παραδοσιακή παράμετρος	parâmetro convencional	конвенциональный параметр	parámetro convencional
κυρτή συνάρτηση	função convexa	выпуклая функция	función convexa
συνέλιξη	convolução	конволюция	convolución
συσχετισμός	corregionalização	совместное локализирование	corregionalización
συσχέτιση	correlação	корреляция	correlación
συντελεστής συσχέτισης	coeficiente de correlação	коэффициент корреляции	coeficiente de correlación
συσχετόγραμμα	correlograma	коррелограмма	correlograma
συνπροσομοίωση	cossimulação	совместное моделирование, соответствующее моделирование	cosimulación
συνδιασπορά	covariância	ковариация	covarianza
πίνακας συνδιασποράς	matriz de covariância	матрица ковариации	matriz de covarianza
εγκάρσια συσχέτιση, διασυσχέτιση	correlação cruzada	взаимная корреляция	correlación cruzada
συντελεστής εγκάρσιας συσχέτισης	coeficiente de correlação cruzada	коэффициент взаимной корреляции	coeficiente de correlación cruzada
συσχετόγραμμα	correlograma cruzado	коррелограмма взаимной корреляции	correlograma cruzado

English	Chinese	French	German
cross-covariance	互协方差	covariance croisée	Kreuzkovarianz
cross-plot	散点图	nuage de corrélation	Kreuzdiagramm
cross-semivariogram	互半变差函数(图)	demi-variogramme croisé	Kreuzsemivariogramm
cross-validation	交叉验证	validation croisée	Kreuzprüfung
cross-variogram	互变差函数(图)	variogramme croisé	Kreuzvariogramm
cubic semivariogram	立方半变差函数(图)	variogramme cubique	kubisches Semivariogramm
cumulative distribution function	累积分布函数	fonction de répartition	Verteilungsfunktion
cumulative frequency	累积频率	frequence cumulée	kumulative Häufigkeit, Summenhäufigkeit
curve fitting	曲线拟合	ajustement de courbe	Kurvenanpassung
data	数据	données	Daten
datum	数；基准面	donnée	a) Bezugsgröße, b) Datum
decile	十分位数	decile	Dezil
declustering	减弱丛聚	dégroupement	Entclusterung
deconvolution	反卷积	déconvolution	Entfaltung
Delaunay subdivision	德劳内剖分	partition de Delaunay	Delaunay-Zerlegung
Delaunay triangulation	德劳内三角剖分	triangulation de Delaunay	Delaunay-Triangulation
demivariogram	半变差函数(图)	demi-variogramme	Semivariogramm
determinant	行列式	déterminant	Determinante
determinant of a matrix	矩阵的行列式	déterminant d'une matrice	Determinante einer Matrix
deterministic geostatistics	确定性地质统计学	géostatistique déterministe	deterministische Geostatistik
de Wijs formula	德威依斯公式	formule de de Wijs	de Wijssche Formel
de Wijsian semivariogram	德威依斯半变差函数(图)	schéma de De Wijs	de Wijssches Semivariogramm
diagonal bivariate density function	对角二元密度函数	densité bivariable isofactorielle	isofaktorielle zweidimensionale Dichte
diagonalized matrix	对角化矩阵	matrice diagonalisée	diagonalisierte Matrix
diagonal matrix	对角矩阵	matrice diagonale	Diagonalmatrix
dilution random function	淡化随机函数	schéma de dilution	Verdünnungs-Zufallsvariable
Dirac delta function	狄拉克 δ 函数	mesure de Dirac	Diracsche Delta-Funktion
direct problem	正问题	problème direct	direkte Aufgabe
Dirichlet tesselation	狄利克雷嵌形	mosaique de Dirichlet	Dirichlet-Parkettierung
discrete Gaussian model	离散高斯模型	modèle gaussien discrétisé	diskretes Gaußsches Modell
discrete random variable	离散随机变量	variable aléatoire discrète	diskrete Zufallsvariable
disjunctive kriging	析取克里格法	krigeage disjonctif	disjunktives Krigen
dispersion variance	离散方差	variance de dispersion	Dispersionsvarianz
distribution	分布	distribution	Verteilung
distribution function	分布函数	fonction de répartition	Verteilungsfunktion
domain	域	champ, domaine	Domäne, Feld
drift	漂移	derive	Drift
dual kriging	对偶克里格法	krigeage dual	duales Krigen
eigenvalue	特征值	valeur propre	Eigenwert

Greek	Portuguese	Russian	Spanish
εγκάρσια συνδιασπορά	covariância cruzada	взаимная ковариация	covarianza cruzada
εγκάρσιο διάγραμμα	núvem de correlação	взаимное корреляционное поле	diagrama de dispersión
εγκάρσιο ημιβαριόγραμμα	semivariograma cruzado	взаимная семивариограмма	semivariograma cruzado
αντεπικύρωση	validação cruzada	взаимная обоснованность	validación cruzada
εγκάρσιο βαριόγραμμα	variograma cruzado	взаимная вариограмма	variograma cruzado
κυβικό ημιβαριόγραμμα	variograma cúbico	кубическая семивариограмма	semivariograma cúbico
αθροιστική συνάρτηση κατανομής	função de distribuição cumulativa	интервальная функция распределения	función de distribución acumulada
αθροιστική συχνότης	frequência acumulada	накопленная частота	frecuencia acumulada
προσαρμογή καμπύλης	ajustamento de curvas	аппроксимация кривой	ajuste de una curva
δεδομένα	dados	данные	datos
επίπεδο αναφοράς	a) datum, b) dado, amostra	данная величина	a) datum, b) dato
δεκατημόριο	decil	дециль	decil
αποσυσσώρευση	desaglomeração	дегруппирование	desaglomeración
αποσυνέλιξη	desconvolução	де-конволюция	deconvolución
υποδιαίρεση Delaunay	subdivisão de Delaunay	подразделение Делона	subdivisión de Delaunay
τριγωνισμός Delaunay	triangulação de Delaunay	триангуляция Делона	triangulación de Delaunay
ημιβαριόγραμμα	semivariograma	демивариограмма	semivariograma
ορίζουσα	determinante	детерминант, определитель	determinante
ορίζουσα ενός πίνακα	determinante de uma matriz	определитель матрицы	determinante de una matriz
προσδιοριστική γεωστατιστική	Geoestatística determinista	детерминистская геостатистика	geoestadística determinística
τύπος του de Wijs	fórmula de de Wijs	формула де Вийса	fórmula de de Wijs
ημιβαριόγραμμα de Wijs	variograma de de Wijs, esquema de de Wijs	семивариограмма де Вийса	semivariograma de de Wijs
διαγώνιος διμεταβλητή συνάρτηση πυκνότητος	função densidade de probabilidade bidimensional isofactorial	функция плотности диагональной двумерной переменной	función de densidad diagonal bivariable
διαγωνοποιημένο μητρώο	matriz diagonalizada	диагонализируемая матрица	matriz diagonalizada
διαγώνιος πίνακας	matriz diagonal	диагональная матрица	matriz diagonal
τυχαία συνάρτηση αραίωσης	função aleatória de diluição	случайная функция разбавления	función aleatoria de dilución
δέλτα Dirac	função delta de Dirac, medida de Dirac	дельта-функция Дирака	función delta de Dirac
άμεσο πρόβλημα	problema directo	прямая задача	problema directo
μωσαϊκό Dirichlet	tecelagem de Dirichlet	замощение Дирихле	teselado de Dirichlet
διακεκριμμένο μοντέλο Gauss	modelo gaussiano discretizado	дискретная гауссовская модель	modelo gaussiano discreto
διακεκριμμένη τυχαία μεταβλητή	variável aleatória discreta	дискретная случайная переменная	variable aleatoria discreta
διαζευτικό κρίκιγκ	krigagem disjuntiva	дизъюнктивный крайгинг	krigeage disyuntivo
διασπορά σκέδασης	variância de dispersão	дисперсия среднего квадратического отклонения	varianza de dispersión
κατανομή	distribuição	распределение	distribución
συνάρτηση κατανομής	função de distribuição	функция распределения	función de distribución
πεδίο ορισμού	campo, domínio	область	dominio
τάση	deriva, tendência	дрифт	tendencia
δυϊκό κρίκιγκ	krigagem dual	двойственный крайгинг	krigeage dual
ιδιοτιμή, χαρακτηριστική τιμή	valor próprio	собственное значение	valor propio, autovalor

English	Chinese	French	German
eigenvalue of a matrix	矩阵的特征值	valeur propre d'une matrice	Eigenwert einer Matrix
eigenvector	特征向量	vecteur propre	Eigenvektor
element of a determinant	行列式的元素	élément d'un déterminant	Element in einer Determinante
element of a matrix	矩阵的元素	élément d'une matrice	Element einer Matrix
elliptical anisotropy	椭园各向异性	anisotropie elliptique	elliptische Anisotropie
ensemble	总体	ensemble	Ensemble, Gesamtheit
entropy	熵	entropie	Entropie
ergodicity	遍历性	ergodicité	Ergodizität
ergodic property	遍历性质	propriété ergodique	Ergodeneigenschaft
ergodic theorem	遍历定理	théorème ergodique	Ergodensatz
error	误差	erreur	Fehler
estimate	估计值	valeur estimée	Schätzwert
estimation	估计	estimation	Schätzung
estimation variance	估计方差	variance d'estimation	Schätzvarianz
estimator	估计量	estimateur	Schätzer
E-type estimate	E 型估计值	estimateur de type E	Schätzwert vom E-typ
euclidean space	欧几里德空间	espace euclidien	Euklidischer Raum
event	事件	évènement	Ereignis
event space	事件空间	espace des échantillons	Ereignisraum
exact interpolator	精确内插	interpolateur exacte	Interpolator im engeren Sinne
expected value	期望值	espérance mathématique	Erwartungswert
experiment	实验	expérience	Versuch, Experiment
experimental covariance	实验协方差	covariance expérimentale	experimentelle Kovarianz
experimental semivariogram	实验半变差函数(图)	variogramme expérimental	experimentelles Semivariogramm
exponential distribution	指数分布	loi exponentielle	Exponentialverteilung
exponential semivariogram	指数半变差函数(图)	variogramme exponentiel	exponentielles Semivariogramm
extension variance	外延方差	variance d'extension	Ausdehnungsvarainz
external drift method	外漂移法	méthode de la dérive externe	Methode der externen Drift
extrapolation	外推	extrapolation	Extrapolation
factorable random function	可因子化随机函数	fonction aléatoire factorisable	faktorisierbare Zufallsfunktion
factorable random process	可因子化随机过程	processus aléatoire factorisable	faktorisierbarer Zufallsprozeß
factor analysis	因子分析	analyse factorielle	Faktorenanalyse
factorial kriging	因子克里格法	analyse krigeante	faktorielles Krigen
fast Fourier transform	快速傅里叶变换	transformation de Fourier rapide	schnelle Fouriertransformierte
F-function	F 函数	variogramme moyen	F-Funktion
field	场	champ	Feld
filtering	滤波	filtrage	filtern
forward problem	正问题	problème direct	direkte Aufgabe
Fourier analysis	傅里叶分析	analyse de Fourier	Fourieranalyse

Greek	Portuguese	Russian	Spanish
ιδιοτιμή ενός πίνακα	valor próprio de uma matriz	собственное значение матрицы	valor propio de una matriz, autovalor de una matriz
χαρακτηριστικό διάνυσμα	vector próprio	собственный вектор	vector propio, autovector
στοιχείο μιας ορίζουσας	elemento de um determinante	элемент детерминанта	elemento de un determinante
στοιχείο ενός πίνακα	elemento de uma matriz	элемент матрицы	elemento de una matriz
ελλειπτική ανισοτροπία	anisotropia elíptica	эллиптическая анизотропия	anisotropía elíptica
σύνολο	conjunto	ансамбль	conjunto
εντροπία	entropia	энтропия	entropía
εργοδικότης	ergodicidade	эргодичность, эргодичество	ergodicidad
εργοδική ιδιότης	propriedade ergódica	свойство эргодичности	propiedad ergódica
εργοδικό θεώρημα	teorema ergódico	эргодическая теорема	teorema ergódico
λάθος	erro	ошибка	error
εκτίμηση	estimador, valor estimado	статистическая оценка	valor estimado
εκτίμηση	estimação	оценка, оценивание	estimación
διασπορά εκτίμησης	variância de estimação	оценка дисперсии	varianza de estimación
εκτιμητής	estimador	статистика	estimador
εκτιμητής τύπου Ε	estimador de tipo-E	оценка типа э	estimador de tipo E
Ευκλίδειος χώρος	espaço euclidiano	эвклидово пространство	espacio euclidiano
ενδεχόμενο, συμβάν	acontecimento	событие	evento, suceso
χώρος ενδεχομένων	espaço dos acontecimentos	пространство выборок	espacio muestral
ακριβής παρεμβολή	interpolador exacto	точный интерполятор	interpolador exacto
αναμενόμενη τιμή	esperança matemática, valor esperado	математическое ожидание	valor esperado
πείραμα	experiência	эксперимент	experimento
πειραματική συνδιασπορά	covariância experimental	экспериментальная ковариация	covarianza experimental
πειραματικό ημιβαριόγραμμα	variograma experimental	экспериментальная семивариограмма	semivariograma experimental
εκθετική κατανομή	distribuição exponencial	экспоненциальное распределение	distribución exponencial
εκθετικό ημιβαριόγραμμα	variograma exponencial	экспоненциальная семивариограмма	semivariograma exponencial
διασπορά επέκτασης	variância de extensão	дисперсия распространения	varianza de extensión
μέθοδος εξωτερικής τάσης	método da deriva externa	метод внешнего дрейфа	método de la tendencia externa
επέκταση	extrapolação	экстраполяция, экстраполирование	extrapolación
παραγοντική τυχαία συνάρτηση	função aleatória factorizável	факторизуемая случайная функция	función aleatoria factorizable
παραγοντική τυχαία διαδικασία	processo aleatório factorizável	факторизуемый случайный процесс	proceso aleatorio factorizable
ανάλυση παραγόντων	análise factorial	факторный анализ	análisis factorial
παραγοντικό κρίκιγκ	krigagem factorial	факториальный крайгинг	krigeage factorial
ταχύς μετασχηματισμός Fourier	transformada rápida de Fourier	быстрое преобразование Фурье	transformada rápida de Fourier
συνάρτηση F	função F	ф-функция, функция ф	función F
πεδίο	campo	поле	campo
φιλτράρισμα	filtragem	фильтрация, фильтрирование	filtrado
προδρομικό πρόβλημα	problema directo	синхронная задача	problema directo
ανάλυση Fourier	análise de Fourier	анализ Фурье	análisis de Fourier

English	Chinese	French	German
Fourier synthesis	傅里叶合成	transformation de Fourier inverse	Fouriersynthese
Fourier transform	傅里叶变换	transformation de Fourier	Fouriertransformierte
fractal	分形	fractal	Fraktal
fractal dimension	分数维	dimension fractale	fraktale Dimension
fractile	分位数	fractile quantile	Fraktil
frequency curve	频率曲线	courbe de fréquence	Häufigkeitskurve
frequency diagram	频率图	diagramme de frequence	Häufigkeitsdiagram
frequency function	频率函数	fonction de fréquence	Häufigkeitsfunktion
function	函数	fonction	Funktion
fuzzy indicator data	模糊指标数据	donneé indicatrice floue	Fuzzy Indikatordaten
gamma distribution	γ分布	loi Gamma	Gammaverteilung
gamma function	γ函数	fonction Gamma	Gammafunktion
Gaussian anamorphosis	高斯变形	anamorphose gaussienne	Gaußscher Anamorphismus
Gaussian distribution	高斯分布	loi de Gauss	Gausßche Verteilung
Gaussian semivario-gram	高斯半变差函数(图)	variogramme gaussien	Gaußsches Semivario-gramm
generalized covariance of order ν	ν 阶广义协方差	covariance generalisée d'ordre ν	verallgemeinerte Kovari-anzen der Ordnung ν
generalized increments of order ν	ν 阶广义增量	increment généralisé d'ordre ν	verallgemeinerte Inkre-mente der Ordnung ν
generalized semivario-gram	广义半变差函数(图)	variogramme généralisé	verallgemeinertes Semivariogramm
general relative semivar-iogram	一般相对半变差函数(图)	demi-variogramme relatif general	allgemeines relatives Semivariogramm
geometric anisotropy	几何各向异性	anisotropie géométrique	geometrische Anisotropie
geometric average	几何平均值	moyenne géométrique	geometrisches Mittel
geometric covariogram	几何协变差函数	covariogramme géo-métrique	geometrisches Kovario-gramm
geometric field	几何场	champ géométrique	geometrisches Feld
geometric mean	几何平均值	moyenne géométrique	geometrisches Mittel
geometric support	几何承载	support géométrique	geometrische Stützung
geostatistics	地质统计学	géostatistique	Geostatistik
global	整体的	global	global
global estimation	整体估计	estimation globale	globale Schätzung
graded semivariogram	级配半变差函数(图)	variogramme monté	gradiertes Semivario-gramm
grading	级配	montée	gradieren
grid	网格	grille	Gitter
hard data	硬数据	donnée dure	harte Daten
harmonic	调和的	harmonique	harmonisch
harmonic mean	调和平均值	moyenne harmonique	harmonisches Mittel
Hausdorff–Besicovitch dimension	豪斯道夫-贝西科维奇维	dimension de Hausdorff–Besicovitch	Hausdorff–Besicovitch-Dimension
Hausdorff measure	豪斯道夫测度	mesure de Hausdorff	Hausdorff-Maß
Heaviside's function	海维赛德函数	fonction d'Heaviside	Heavisidesche Funktion
Hermite model	埃尔米特模型	modèle hermitien	Hermitesches Modell

Greek	Portuguese	Russian	Spanish
σύνθεση Fourier	transformada de Fourier inversa	синтез Фурье	síntesis de Fourier
μετασχηματισμός Fourier	transformada de Fourier	преобразование Фурье	transformada de Fourier
τομικός	fractal	фрактал, множество дробной размерности	fractal
κλασματική διάσταση	dimensão fractal	дробная размерность	dimensión fractal
ποσοστιαίο σημείο	quantil	квантиль	fractil
καμπύλη συχνότητος	curva de frequências	кривая частоты	curva de frecuencias
διάγραμμα συχνότητος	diagrama de frequências	диаграмма частоты	diagrama de frecuencias
συνάρτηση συχνότητας	função frequência	функция частоты	función de frecuencia
συνάρτηση	função	функция	función
θαμπά δεικτικά δεδομένα	dados vagos de indicatriz	метка размытого множества	datos de un indicador vago
κατανομή γάμμα	distribuição Gama	гамма-распределение	distribución gamma
συνάρτηση γάμμα	função gama	гамма-функция	función gamma
αναμόρφωση Gauss	anamorfose gaussiana	анаморфоз Гаусса	anamorfosis gaussiana
κατανομή Gauss	distribuição gaussiana	распределение Гаусса, нормальное гауссовское распределение	distribución de Gauss
ημιβαριόγραμμα Gauss	variograma gaussiano	семивариограмма Гаусса	semivariograma de Gauss
γενικευμένη συνδιασπορά τάξης v	covariância generalizada de ordem v	обобщенная ковариация порядка v	covarianza generalizada de orden v
γενικευμένες αυξήσεις τάξης v	incrementos generalizados de ordem v	обобщенные приросты порядка v	incrementos generalizados de orden v
γενικευμένο γραμμικό ημιβαριόγραμμα	variograma generalizado	обобщенная семивариограмма	semivariograma generalizado
γενικό σχετικό ημιβαριόγραμμα	variograma global relativo	генеральная относительная семивариограмма	semivariograma relativo general
γεωμετρική ανισοτροπία	anisotropia geométrica	геометрическая анизотропия	anisotropía geométrica
γεωμετρικός μέσος όρος	média geométrica	геометрическое среднее	promedio geométrico
γεωμετρικό συνδιασπορόγραμμα	covariograma geométrico	геометрическая ковариограмма	covariograma geométrico
γεωμετρικό πεδίο	campo geométrico	геометрическое поле	campo geométrico
γεωμετρική μέση τιμή	média geométrica	геометрическая средняя, среднее геометрическое	media geométrica
γεωμετρική υποστήριξη	suporte geométrico	геометрический объем	soporte geométrico
γεωστατιστική	Geoestatística	геостатистика	geoestadística
ολικός	global	глобальный	global
ολική εκτίμηση	estimação global	глобальная оценка	estimación global
βαθμωτό ημιβαριόγραμμα	variograma acumulado	градуированная семивариограмма	semivariograma acumulado
διαβάθμιση	acumulação	градуирование	acumulación
εσχάρα	malha, grelha	сеть	cuadrícula, red, malla, grilla, reticulado
σκληρά δεδομένα	dados fiáveis	точные данные	datos fiables
αρμονικός	harmónica	гармонический	armónica
αρμονική μέση τιμή	média harmónica	гармоническая средняя, среднее гармоническое	media armónica
διάσταση Hausdorff–Besicovitch	dimensão de Hausdorff-Besicovitch	размерность Гаусдорфа-Безиковича	dimensión de Hausdorff–Besicovitch
μέτρο Hausdorff	medida de Hausdorff	мера Гаусдорфа	medida de Hausdorff
συνάρτηση Heaviside	função de Heaviside	единичная функция Хевисайда, функция Хевисайда	función de Heaviside
μοντέλο Hermite	modelo hermitiano	модель Эрмита	modelo de Hermite

English	Chinese	French	German
Hermite polynomials	埃尔米特多项式	polynômes d'Hermite	Hermitesche Polynome
Hermitian matrix	埃尔米特矩阵	matrice Hermitienne	Hermitesche Matrix
heteroscedastic	异方差的	hétéroscédastique	heteroskedastisch
heterotopic	异位的	hétérotopique	heterotop
high-grade destructuration	高品位去结构化	déstructuration des hautes teneurs	Destrukturierung großer Gehalte
high-grade structuration	高品位结构化	structuration des hautes teneurs	Strukturierung hoher Gehalte
Hilbert space	希尔伯特空间	espace de Hilbert	Hilbert-Raum
histogram	直方图	histogramme	Histogramm
hole effect	空穴效应	effet de trou	Locheffeekt
hole effect semivariogram model	空穴效应半变差函数模型	modèle de variogramme à effet de trou	Semivariogrammodell mit Locheffeekt
homogeneity	齐性	homogénéité	Homogenität
homoscedastic	同方差的	homoscedastique	homoskedastisch
h-scattergram	h 散点图	nuage de corrélation différée	h-Streudiagramm
identity matrix	单位矩阵	matrice identité	Einheitsmatrix
impulse	脉冲	impulsion	Impuls
indicator	指标	indicatrice	Indikator
indicator cokriging	指标协克里格法	cokrigeage d'indicatrices	Indikatorkokrigen
indicator covariance	指标协方差	covariance d'indicatrice	Indikatorkovarianz
indicator cross-covariance	指标互协方差	covariance d'indicatrice croisée	Kreuzindikatorkovarianz
indicator data column	指标数据列	colonne des données d'indicatrice	Indikatordatenspalte
indicator kriging	指标克里格法	krigeage d'indicatrice	Indikatorkrigen
indicator principal component kriging	指标主分量克里格法	krigeage des composantes principales d'indicatrices	Krigen der Indikatorhauptkomponenten
indirect lognormal correction	间接对数校正	correction lognormale indirecte	indirekte lognormale Korrektur
infimum	下确界	infimum	Infimum
infinite divisibility	无限可分性	infinie divisibilité	unbegrenzte Teilbarkeit
infinite divisibility of a union	并集的无限可分性	infinie divisibilité pour la réunion	unbegrenzte Zerlegbarkeit einer zufälligen Vereinigungsmenge
information effect	信息效应	effet d'information	Informationseffekt
inner product	内积	produit scalaire	inneres Produkt
inner product space	内积空间	espace de produit scalaire	Raum, in dem ein inneres Produkt definiert ist
integrable function	可积函数	fonction intégrable	integrierbare Funktion
integrable function with respect to a second function	对于第二个函数的可积函数	fonction intégrable par rapport à une deuxième fonction	integrierbare Funktion in Bezug auf eine zweite Funktion
integral range	积分范围	portée intégral	Integrationsbereich
interior	内部	intérieur	das Innere
intermeshed structure	套合结构	structures gigognes	geschachtelte Struktur
interpolation	内插	interpolation	Interpolation
interquartile range	四分位数间距	interquartile	Intervall zwischen dem unteren und oberen Quartil

Greek	Portuguese	Russian	Spanish
πολυώνυμα Hermite	polinómios de Hermite	многочлены Эрмита	polinomios de Hermite
μητρώο Hermite	matriz Hermitiana	матрица Эрмита	matriz de Hermite
ετεροσκεδαστικότης	heteroscedástico	гетероскедастический	heteroscedástico
ετεροτοπικός	heterotópico	гетеротопический	heterotópico
αποδόμηση υψηλών περιεκτικοτήτων	desestruturação dos altos teores	деструктуризация высокого содержания	desestructuración de las leyes altas
προσδιορισμός υψηλών περιεκτικοτήτων	estruturação dos altos teores	структуризация высокого содержания	estructuración de las leyes altas
χώρος Hilbert	espaço de Hilbert	Гильбертово пространство	espacio de Hilbert
ιστόγραμμα	histograma	гистограмма	histograma
επίδραση οπής	efeito de cabeça de cavalo	дырочный эффект	efecto agujero
ημιβαριόγραμμα επίδρασης οπής	modelo de variograma com efeito de cabeça de cavalo	модель дырочного эффекта семивариограммы	semivariograma de efecto agujero
ομοιογένεια	homogeneidade	гомогенность, однородность	homogeneidad
ομοσκεδαστικότης	homoscedástico	гомоскедастический	homoscedástico
h-διασπορόγραμμα	núvem de correlação a passo h	г-диаграмма рассеивания	diagrama de dispersión tipo h
ταυτοτικός πίνακας	matriz identidade	единичная матрица	matriz identidad
παλμός	impulso	дельта-функция Дирака	impulso
δείκτης	indicatriz	индикатор	indicador
κοκρίκιγκ δείκτη	cokrigagem da indicatriz	индикатор кокрайгинга	cokrigeage de indicadores
συνδιασπορά δείκτη	covariância da indicatriz	индикатор ковариация	covarianza de indicadores
εγκάρσια συνδιασπορά δείκτη	covariância cruzada da indicatriz	взаимная ковариация индикатора	covarianza cruzada de indicadores
στήλη δεδομένων δείκτη	coluna dos dados da indicatriz	столбец данных индикатора	columna de datos de un indicador
κρίκιγκ δείκτη	krigagem da indicatriz	индикатор крайгинг	krigeage de un indicador
κρίκιγκ δείκτη κυρίων συνιστωσών	krigagem das componentes principais da indicatriz	крайгинг главных компонентов индикатора	krigeage de las componentes principales de los indicadores
έμμεση λογοκανονική διόρθωση	correcção lognormal indirecta	непрямая логнормальная коррекция	corrección lognormal indirecta
κατώτατο όριο	ínfimo	инфимум	ínfimo
άπειρος διαιρετότης	infinitamente divisível	бесконечная делимость	divisibilidad infinita
άπειρος διαιρετότης μιάς ένωσης	infinitamente divisível para a união	бесконечная делимость объединения	divisibilidad infinita de una unión
επίδραση πληροφόρησης	efeito de informação	информационный эффект	efecto información
εσωτερικό γινόμενο	produto interno	внутреннее произведение	producto interior
χώρος εσωτερικού γινομένου	espaço vectorial munido de produto interno	пространство с операцией внутреннего произведения	espacio producto interior
ολοκληρώσιμη συνάρτηση	função integrável	интегрируемая функция	función integrable
ολοκληρώσιμη συνάρτηση ως προς μια δεύτερη συνάρτηση	função integrável em relação a uma segunda função	интегрируемая функция по второй производной функции	función integrable con respecto a una segunda función
διάστημα ολοκλήρωσης	alcance integral	интегральная область значений	alcance integral
εσωτερικός	interior	внутренний	interior
εσωδικτυακή κατασκευή	estrutura embricada	гнездовая структура	estructura imbricada, estructura compuesta
παρεμβολή	interpolação	интерполяция	interpolación
διάστημα ενδοτεταρτημορίου	intervalo inter-quartílico	интерквартильная широта, расстояние между квартилями	amplitud intercuartílica

English	Chinese	French	German
intrinsic	内蕴的	intrinsèque	intrinsisch
intrinsic attribute	内蕴属性	caractéristique intrinsèque	intrinsisches Attribut
intrinsic correlation	内蕴相关	corrélation intrinsèque	intrinsische Korrelation
intrinsic dispersion	内蕴离散	dispersion intrinsèque	intrinsische Dispersion
intrinsic function	内蕴函数	fonction intrinsèque	intrinsische Funktion
intrinsic hypothesis	内蕴假设	hypothèse intrisèque	intrinsische Hypothese
intrinsic kriging	内蕴克里格法	krigeage intrinsèque	intrinsisches Krigen
intrinsic random function of order ν	ν阶内蕴随机函数	fonction aléatoire intrinsèque d'ordre n	intrinsische Zufallsfunktion der Ordnung ν
inverse of a matrix	矩阵的逆	matrice inverse	Inverse einer Matrix
inverse problem	逆问题	problème inverse	inverse Aufgabe
isofactorial bivariate density function	等因子二元密度函数	densité bivariable isofactorielle	isofaktorielle zweidimensionale Dichte
isoline	等值线	courbe de niveau	Isolinie
isopleth	等值线	courbe de niveau	Isoplethe, Linie gleicher Werte
isotopic	同位的	isotopique	isotop
isotropy	各向同性	isotropie	Isotropie
jackknife	折刀法	Jackknife	Jackknife
joint cross-validation	联合交叉验证	validation croisée multiple	gemeinsame Kreuzprüfung
jointly factorable random function	联合可因子化随机函数	fonction aléatoire simultanément factorisable	gemeinsam faktorisierbare Zufallsfunktion
jointly orthogonal random function	联合正交随机函数	fonction aléatoire simultanément orthogonal	gemeinsam orthogonale Zufallsfunktion
jointly orthogonal random process	联合正交随机过程	processus aléatoire simultanément orthogonal	gemeinsam orthogonaler Zufallsprozeß
joint probability density function	联合概率密度函数	densité multivariable	gemeinsame Wahrscheinlichkeitsdichte
joint probability distribution	联合概率分布	loi multivariable	gemeinsame Wahrscheinlichkeitsfunktion
Krige's ellipse	克里格椭园	ellipse de Krige	Krigesche Ellipse
Krige's formula	克里格公式	formule de Krige	Krige-Beziehung
kriging	克里格法	krigeage	Krigen
kriging neighborhood	克里格邻域	voisinage de krigeage	Krigenachbarschaft
kriging variance	克里格方差	variance de krigeage	Krigevarianz
Kronecker delta	克罗内克符号	symbole de Kronecker	Kroneckersches delta
kurtosis	峰度	aplatissement	Wölbung
lag	滞后距	retard	Schrittweite, Abstand
Lagrange's method of multipliers	拉格朗日乘子法	formalisme de Lagrange	Lagrangesche Multiplikatoren Methode
Lagrange multiplier	拉格朗日乘子	multiplicateur de Lagrange	Lagrangesche Multiplikatoren
Laguerre polynomials	拉盖尔多项式	polynôme de Laguerre	Laguerresche Polynome
Laplace transform	拉普拉斯变换	transformation de Laplace	Laplacesche Transformierte
law of large numbers	大数定律	loi des grands nombres	Gesetz der großen Zahlen
least squares method	最小二乘法	méthode des moindres carrés	Methode der kleinsten Quadrate
leptokurtic	尖峰态	leptokurtique	steilgipflig
likelihood function	似然函数	fonction de vraisemblance	Likelihood-Funktion

Greek	Portuguese	Russian	Spanish
εγγενές	intrínseco	внутренний	intrínseco
εγγενές χαρακτηριστικό	característica intrínseca	внутреннее свойство, внутренний признак	atributo intrínseco
εγγενής συσχέτιση	correlação intrínseca	внутренняя корреляция	correlación intrínseca
εγγενής διασπορά	dispersão intrínseca	внутренняя дисперсия	dispersión intrínseca
εγγενής συνάρτηση	função intrínseca	внутренняя функция	función intrínseca
εγγενής υπόθεση	hipótese intrínseca	внутренняя гипотеза	hipótesis intrínseca
εγγενές κρίκιγκ	krigagem intrínseca	внутренний крайгинг	krigeage intrínseco
εγγενής τυχαία συνάρτηση τάξης ν	função aleatória intrínseca de ordem ν	внутренняя случайная функция порядка ν	función aleatoria intrínseca de orden ν
αντίστροφος ενός πίνακα	matriz inversa	обращение матрицы	matriz inversa
αντίστροφο πρόβλημα	problema inverso	обратная задача	problema inverso
ισοπαραγοντική διμεταβλητή συνάρτηση πυκνότητας	função densidade de probabilidade bidimensional isofactorial	функция плотности изофакториальной двумерной случайной переменной	función de densidad isofactorial bivariable
ισογραμμή	isolinha	изолиния	isolínea
ισοπληθής	isolinha	изоплета	isopleta
ισοτοπικό	isotópica	изотопический	isotópico
ισοτροπία	isotropia	изотропия	isotropía
jackknife	técnica de jackknife	оценка складного ножа, оценка "jackknife"	jackknife
από κοινού αντεπικύρωση	validação cruzada conjunta	совместная взаимная обоснованность	validación cruzada conjunta
από κοινού παραγοντικές τυχαίες συναρτήσεις	funções aleatórias conjuntamente factorizáveis	совместно факторизуемая случайная функция	función aleatoria simultáneamente factorizable
από κοινού ορθογωνικές τυχαίες συναρτήσεις	função aleatória conjuntamente ortogonal	совместно ортогональная случайная функция	función aleatoria simultáneamente ortogonal
από κοινού ορθογωνικές τυχαίες διαδικασίες	processo aleatório conjuntamente ortogonal	совместно ортогональный случайный процесс	proceso aleatorio simultáneamente ortogonal
από κοινού πυκνότης πιθανότητας	função densidade de probabilidade multidimensional	функция плотности совместной вероятности	función de densidad de probabilidad conjunta
από κοινού κατανομή πιθανότητας	função de distribuição multidimensional	распределение совместной вероятности	función de distribución de probabilidad conjunta
έλλειψη Krige	elipse de Krige	эллипс Криге	elipse de Krige
τύπος Krige	fórmula de Krige	формула Криге	fórmula de Krige
κρίκιγκ	krigagem	крайгинг	krigeage, krigeado
γειτονιά κρίκιγκ	vizinhança de krigagem	окрестность крайгинга	entorno de krigeage
διασπορά κρίκιγκ	variância de krigagem	дисперсия крайгинга	varianza de krigeage
δέλτα Kronecker	símbolo delta de Kronecker	символ Кронекера	delta de Kronecker
κύρτωση	achatamento, curtose	эксцесс	curtosis
υστέρηση	passo, malha	запаздывание	desfasamiento, rezago
μέθοδος των πολλαπλασιαστών Lagrange	método dos multiplicadores de Lagrange	метод неопределенных множителей Лагранжа	método de los multiplicadores de Lagrange
πολλαπλασιαστής Lagrange	multiplicador de Lagrange	множитель Лагранжа	multiplicador de Lagrange
πολυώνυμα Laguerre	polinómios de Laguerre	многочлены Чебышева-Лагерра	polinomios de Laguerre
μετασχηματισμός Laplace	transformada de Laplace	преобразование Лапласа	transformada de Laplace
νόμος των μεγάλων αριθμών	lei dos grandes números	закон больших чисел	ley de los grandes números
μέθοδος των ελαχίστων τετραγώνων	método dos mínimos quadrados	метод наименьших квадратов	método de mínimos cuadrados
λεπτοκυρτικός	leptocúrtica	островершинное	leptocúrtico
συνάρτηση πιθανοφάνειας	função de verosimilhança	функция правдоподобия	función de verosimilitud

English	Chinese	French	German
linear combination of variables	变量的线性组合	combinaison linéaire de variables	Linearkombination von Variablen
linear dependence	线性相关	dependance linéaire	lineare Abhängigkeit
linear filtering	线性滤波	filtrage linéaire	lineares Filtern
linear geostatistics	线性地质统计学	géostatistique linéaire	lineare Geostatistik
linear independence	线性无关	indépendance linéaire	lineare Unabhängigkeit
linear manifold	线性流形	variété linéaire	lineare Mannigfaltigkeit
linear model of coregionalization	协区域化的线性模型	modèle linéaire de corégionalisation	lineares Modell der Koregionalisierung
linear semivariogram	线性半变差函数(图)	variogramme linéaire	lineares Semivariogramm
linear space	线性空间	espace vectoriel	linearer Raum
local	局部的	local	lokal
local estimation	局部估计	estimation locale	lokale Schätzung
local relative semivariogram	局部相对半变差函数(图)	demi-variogramme relatif local	lokales relatives Variogramm
logarithmic kriging	对数克里格法	krigeage logarithmique	logarithmisches Krigen
logarithmic semivariogram	对数半变差函数(图)	variogramme logarithmique	logarithmisches Semivariogramm
lognormal correction	对数正态校正	correction lognormale	lognormale Korrektur
lognormal distribution	对数正态分布	loi lognormale	lognormale Verteilung
lognormal kriging	对数正态克里格法	krigeage lognormal	lognormales Krigen
loss function	损失函数	fonction de perte	Verlustfunktion
lower quartile	下四分位数	quartile inférieur	unteres Quartil
lower-upper (LU) decomposition	LU分解法	décomposition L.U.	Zerlegung in untere und obere Dreiecksmatrix
madogram	一阶变差函数	madogramme	Madogramm
main diagonal of a matrix	矩阵的主对角线	diagonale principale d'une matrice	Hauptdiagonale einer Matrix
manifold	流形	variété	Mannigfaltigkeit
marginal distribution	边缘分布	distribution marginale	Randverteilung
marginal probability density function	边缘概率密度函数	densité marginale	randliche Wahrscheinlichkeitsdichte
Markov process	马尔可夫过程	processus de Markov	Markovscher Prozeß
matrix	矩阵	matrice	Matrix
maximum entropy formalism	极大熵形式体系	formalisme du maximum entropie	Methode der maximalen Entropie
maximum likelihood estimator	极大似然估计量	estimateur du maximum de vraisemblance	Maximum-Likelihood-Schätzer
maximum likelihood statistic	极大似然统计量	estimateur du maximum de vraisemblance	Maximum-Likelihood-Statistik
mean	平均值；均值	espérance mathématique	Mittelwert
mean semivariogram	平均半变差函数	variogramme moyen	mittleres Semivariogramm
mean square error	均方误差	erreur quadratique moyenne	mittlerer quadratischer Fehler
measure	测度	mesure	Maß
median	中位数	médiane	Zentralwert

Greek	Portuguese	Russian	Spanish
γραμμικός συνδιασμός μεταβλητών	combinação linear de variáveis	линейная комбинация переменных	combinación lineal de variables
γραμμική εξάρτηση	dependência linear	линейная зависимость	dependencia lineal
γραμμικό φιλτράρισμα	filtragem linear	линейная фильтрация	filtro lineal
γραμμική γεωστατιστική	Geoestatística linear	линейная геостатистика	geoestadística lineal
γραμμική ανεξαρτησία	independência linear	линейная независимость	independencia lineal
γραμμική πολλαπλότης	variedade linear, subespaço afim	линейное многообразие, векторное подпространство	variedad lineal
γραμμικό μοντέλο συσχέτισης	modelo linear de corregionalização	соответствующая линейная модель соответствующего локализирования	modelo lineal de corregionalización
γραμμικό ημιβαριόγραμμα	variograma linear	линейная семивариограмма	semivariograma lineal
γραμμικός χώρος	espaço vectorial	линейное пространство	espacio lineal
τοπικός	local	локальный	local
τοπική εκτίμηση	estimação local	локальная оценка	estimación local
τοπικό σχετικό ημιβαριόγραμμα	variograma local relativo	локальная относительная семивариограмма	semivariograma relativo local
λογαριθμικό κρίκιγκ	krigagem logarítmica	логарифмический крайгинг	krigeage logarítmico
λογαριθμικό ημιβαριόγραμμα	variograma logarítmico	логарифмическая семивариограмма	semivariograma logarítmico
λογοκανονική διόρθωση	correcção lognormal	логнормальная коррекция	corrección lognormal
λογοκανονική κατανομή	distribuição lognormal	логнормальное распределение	distribución lognormal
λογοκανονικό κρίκιγκ	krigagem lognormal	логнормальный крайгинг	krigeage lognormal
συνάρτηση απώλειας	função de penalização	функция-потерь	función de pérdida
κάτω τεταρτημόριο	primeiro quartil	нижний квартиль	cuartil inferior
κατώτερη-ανώτερη αποσύνθεση (ανάλυση)	decomposição L.U.	нижнее-верхнее (НВ) треугольное разложение	factorización LU
μαντόγραμμα	madograma	мадограмма	madograma
κυρία διαγώνιος ενός πίνακα	diagonal principal de uma matriz	главная диагональ матрицы	diagonal principal de una matriz
πολλαπλότης	variedade	разнообразие	variedad
περιθωριακή κατανομή	distribuição marginal	маргинальное распределение	distribución marginal
περιθωριακή πυκνότης πιθανότητος	função densidade de probabilidade marginal	функция плотности маргинальной вероятности	función de densidad marginal de probabilidad
διαδικασία Markov	processo de Markov	марковский процесс	proceso markoviano
μητρώο, πίνακας	matriz	матрица	matriz
διατύπωση μεγίστης εντροπίας	formalismo de máxima entropia	формализм максимальной энтропии	formalismo de máxima entropía
εκτιμητής μεγίστης πιθανοφάνειας	estimador de máxima verosimilhança	оценка максимального правдоподобия	estimador de máxima verosimilitud
στατιστική μεγίστης πιθανοφάνειας	estimador de máxima verosimilhança	статистика максимального правдоподобия	estadístico de máxima verosimilitud
μέση τιμή	esperança matemática	средняя величина	media
μέσο ημιβαριόγραμμα	variograma médio	средняя семивариограмма	semivariograma promedio
μέσο τετραγωνικό λάθος	erro quadrático médio	среднеквадратичная ошибка	error cuadrático medio
μέτρο	medida	мера	medida
διάμεσος	mediana	медиана	mediana

English	Chinese	French	German
median indicator kriging	中位数指标克里格法	krigeage de indicatrice de médiane	Zentralwertindikatorkrigen
Meijering's cell	梅也伦格子	cellule de Meijering	Meijerings Zelle
mesokurtic	常峰态	mésocurtique	normalgipflig
metric space	度量空间	espace métrique	metrischer Raum
microergodicity	微遍历性	microergodicité	Mikroergodizität
minimum error norm estimator	极小误差模估计量	estimateur minimisant la norme de l'error	Schätzer mit minimaler Fehlernorm
minimum mean square error estimator	极小均方误差估计量	estimateur de moindre erreur carrée	Schätzer mit minimalem mittleren quadratischen Fehler
mode	众数	mode	häufigster Wert
model	模型	modèle	Modell
moment	矩	moment	Moment
moment of a distribution	分布的矩	moment d'une loi	Moment einer Verteilung
monoscopic	单目标的	monoscopique	monoskopisch
Monte Carlo method	蒙特卡罗法	méthode de Monte-Carlo	Monte-Carlo-Methode
mosaic model	镶嵌模型	modèle mosaïque	Mosaikmodell
moving linear average	移动线性平均	moyenne mobile	gleitendes Mittel
multigaussian kriging	多元高斯克里格法	krigeage multigaussien	multigaußsches Krigen
multimodal distribution	多峰分布	distribution multimodale	multimodale Verteilung
multinormal distribution	多元正态分布	loi multigaussienne	mehrdimensionale Normalverteilung
multiple indicator kriging	多重指标克里格法	krigeage d'indicatrice multiple	mehrfaches Indikatorkrigen
multivariate	多元的	multivariable	mehrdimensional
multivariate distribution	多元分布	loi multivariable	mehrdimensionale Verteilung
negative binomial distribution	负二项分布	loi binomiale négative	negative Binomialverteilung
neighborhood	邻域	voisinage	Nachbarschaft, Umgebung
nested semivariogram	套合半变差函数(图)	variogramme gigogne	geschachteltes Semivariogramm
nested structure	套合结构	structures gigognes	geschachtelte Struktur
noncentered covariance	非中心协方差	covariance non centrée	nichtzentrierte Kovarianz
nonergodic experimental covariance	非遍历实验协方差	covariance expérimentale non-ergodique	nichtergodische experimentelle Kovarianz
nonlinear geostatistics	非线性地质统计学	géostatistique non linéaire	nichtlineare Geostatistik
nonparametric geostatistics	非参数地质统计学	géostatistique non paramètrique	nichtparametrische Geostatistik
nonsingular system	非奇异系统	système non-singulier	nichtsinguläres System
norm	模	norme	Norm

Greek	Portuguese	Russian	Spanish
κρίκιγκ διαμέσου ενός δείκτη	krigagem da indicatriz da mediana	медианный индикаторный крайгинг	krigeage de la mediana de un indicador
κελί Meijering	célula de Meijering	клетка, ячейка Мейеринга	celda de Meijering
μεσοκυρτικός	mesocúrtica	мезовершинная	mesocúrtico
μετρικός χώρος	espaço métrico	метрическое пространство	espacio métrico
μικροεργοδικότης	microergodicidade	микроэргодичность	microergodicidad
εκτιμητής ελαχίστου μετρικού λάθους	minimização da norma	оценка нормы минимальной ошибки	estimador de mínima norma del error
ελάχιστο μέσο τετραγωνικό λάθος εκτίμησης	melhor estimador em erro quadrático médio	оценка ошибки минимального квадрата средней величины	estimador de mínimo error cuadrático medio
κορυφή	moda	мода	moda
μοντέλο	modelo	модель	modelo
ροπή	momento	момент	momento
ροπή μιάς κατανομής	momento de uma distribuição	момент распределения	momento de una distribución
μονοσκοπική	monoscópico	моноскопический	monoscópico
μέθοδος Monte Carlo	método de Monte Carlo	метод Монте-Карло, метод статистических испытаний	método de Monte Carlo
μωσαϊκό μοντέλο	modelo de mosaico	мозаичная модель	modelo de mosaico
κινητός γραμμικός μέσος	média móvel	скользящее линейное среднее	media lineal móvil
κρίκιγκ πολυκανονικό	krigagem multigaussiana	многогауссовский крайгинг	krigeage multigaussiano
πολύκορφη κατανομή	distribuição plurimodal	многомодальное распределение	distribución multimodal
πολυκανονική κατανομή	distribuição normal multidimensional	многомерное нормальное распределение	distribución multinormal
κρίκιγκ πολλαπλού δείκτη	krigagem múltipla da indicatriz	кратный индикаторный крайгинг	krigeage de varios indicadores
πολυμεταβλητή	multidimensional	многомерная случайная переменная	multivariable
πολυμεταβλητή κατανομή	distribuição multidimensional	многомерное распределение	distribución multivariable
αρνητική διωνυμική κατανομή	distribuição binomial negativa	отрицательное биномиальное распределение	distribución binomial negativa
γειτονιά	vizinhança	окрестность	entorno
εγκλωβιστικό ημιβαριόγραμμα	variograma embricado	гнездовая семивариограмма	semivariograma imbricado, semivariograma compuesto
εγκλωβιστική κατασκευή	estruturas embricadas	гнездовая структура	estructura imbricada, estructura compuesta
μη κεντρική συνδιασπορά	covariância não centrada	нецентрированная ковариация	covarianza no centrada
μη εργοδική πειραματική συνδιασπορά	covariância experimental não ergódica	неэргодическая экспериментальная ковариация	covarianza experimental no ergódica
μη γραμμική γεωστατιστική	Geoestatística não linear	нелинейная геостатистика	geoestadística no lineal
μη παραμετρική γεωστατιστική	Geoestatística não paramétrica	непараметрическая геостатистика	geoestadística no paramétrica
μη ιδιάζον σύστημα	sistema não-singular	регулярная система	sistema no singular
μέτρο	norma	норма	norma

English	Chinese	French	German
normal distribution	正态分布	loi normale	Normalverteilung
normal equations	正规方程组	équations normales	Normalengleichungen
normal problem	正问题	problème direct	direkte Aufgabe
normal score transform	正态得分变换;正态变换	anamorphose	Anamorphismus
nugget effect	块金效应	effet de pépite	Nuggeteffekt
objective function	目标函数	fonction d'efficacité	Zielfunktion
objective meaning	客观涵义	signification objective, sens objectif	objektive Bedeutung
objectivity criterion	客观性准则	critère d'objectivité	Objektivitätskriterium
open	开的	ouvert	offen
operational concept	操作概念	concept opératoire	Operationskonzept
operator	算子	opérateur	Operator
optimal residual	最优剩余	résidu optimal	optimale Residuen
order of a determinant	行列式的阶	ordre d'un déterminant	Ordnung einer Determinante
order of a square matrix	方阵的阶	ordre d'une matrice carrée	Ordnung einer quadratischen Matrix
ordinary kriging	普通克里格法	krigeage ordinaire	normales Krigen
orthogonal function	正交函数	fonction orthogonale	orthogonale Funktion
orthogonal polynomial	正交多项式	polynôme orthogonal	orthogonale Polynome
orthogonal random function	正交随机函数	fonction aléatoire orthogonale	orthogonale Zufallsfunktion
orthogonal random process	正交随机过程	processus aléatoire orthogonale	orthogonaler Zufallsprozeß
orthogonal vector	正交向量	vecteur orthogonal	orthogonale Vektoren
orthonormal function	标准正交函数	fonction orthonormale	orthonormale Funktionen
orthonormal vector	标准正交向量	vecteur orthonormal	orthonormale Vektoren
outcome	结果	réalisation	Realisierung
outlier	特异值	données aberrantes	Ausreißer
overestimation	过高估计	surestimation	Überschätzung
pairwise relative semivariogram	两两相对半变差函数(图)	demi-variogramme relatif par paire	paarweise relatives Semivariogramm
panscopic	泛目标的	panscopique	panskopisch
parabolic behavior	抛物线性态	comportement parabolique	parabolisches Verhalten
parabolic semivariogram	抛物线半变差函数(图)	variogramme parabolique	parabolisches Semivariogramm
Parseval's equation	帕赛维尔方程	formule de Parseval	Parsevalsche Gleichung
partition of a set	集的分割	partition d'un ensemble	Zerlegung einer Menge
Pearson correlation coefficient	皮尔逊相关系数	coefficient de correlation de Pearson	Pearsonscher Korrelationskoeffizient
penta-spherical semivariogram	五次球状半变差函数(图)	variogramme penta-sphérique	quintisch-sphärisches Semivariogramm
percentile	百分位数	percentile	Perzentil
permissibility criteria	允许性准则	critère de cohérence	Zulässigkeitskriterium
permissible coefficients of order v	v阶允许系数	coefficients autorisés d'ordre v	zulässige Koeffizienten der Ordnung v
permissible combinations of order v	v阶允许组合	combinaisons autorisées d'ordre v	zulässige Kombinationen der Ordnung v
pivot	主元	pivot	Pivotmatrixelemente

Greek	Portuguese	Russian	Spanish
κανονική κατανομή	distribuição normal	нормальное распределение	distribución normal
κανονικές εξισώσεις	equações normais	нормальные уравнения	ecuaciones normales
κανονικό πρόβλημα	problema directo	прямая задача	problema normal
αναμόρφωση	anamorfose gaussiana	нормальное преобразование ядра оператора	transformación normal
επίδραση κόκκου	efeito de pepita	эффект самородков	efecto pepita
συνάρτηση αντικείμενο	função objectivo	целевая функция	función objetivo
αντικειμενική έννοια	significado objectivo	целевое значение, целевой смысл	significado objetivo
κριτήριο αντικειμενικότητας	critério de objectividade	критерий еффективности	criterio de objetividad
ανοικτός	aberto	открытый	abierto
επιχειρησιακή έννοια	conceito de operacionalidade	операционное понятие	concepto operativo
τελεστής	operador	оператор	operador
βέλτιστο υπόλοιπο	resíduo optimal	оптимальный остаток	residuo óptimo
τάξη ορίζουσας	ordem de um determinante	порядок определителя	orden de un determinante
τάξη τετραγωνικού πίνακα	ordem de uma matriz quadrada	порядок квадратной матрицы	orden de una matriz cuadrada
σύνηθες κρίκιγκ	krigagem ordinária	обыкновенный крайгинг	krigeage ordinario
ορθογωνική συνάρτηση	função ortogonal	ортогональная функция	función ortogonal
ορθογωνικά πολυώνυμα	polinómio ortogonal	ортогональный многочлен	polinomio ortogonal
ορθογωνική τυχαία συνάρτηση	função aleatória ortogonal	ортогональная случайная функция	función aleatoria ortogonal
ορθογωνική τυχαία διαδικασία	processo aleatório ortogonal	ортогональный случайный процесс	proceso aleatorio ortogonal
ορθογωνικό διάνυσμα	vector ortogonal	ортогональный вектор	vector ortogonal
ορθοκανονική συνάρτηση	funções ortonormais	ортонормальная функция	función ortonormal
ορθοκανονικό διάνυσμα	vectores ortonormados	ортонормальный вектор	vector ortonormal
έκβαση	resultado	реализация	resultado
απομονωμένη τιμή	valor extremo aberrante	выбраковка	valor errático, valor atípico
υπερεκτίμηση	sobreestimação	переоценка переоценивание	sobreestimación
κατά ζεύγη σχετικό ημιβαριόγραμμα	variograma experimental relativo	попарно относительная семивариограмма	semivariograma relativo por parejas
πανσκοπική	panscópico	панскопический	panescópico
παραβολική συμπεριφορά	comportamento parabólico	процесс по параболе	comportamiento parabólico
παραβολικό ημιβαριόγραμμα	variograma parabólico	параболическая семивариограмма	semivariograma parabólico
εξίσωση Parseval	equação de Parseval	равенство Парсеваля	ecuación de Parseval
διαμερισμός ενός συνόλου	partição de um conjunto	разбиение множества	partición de un conjunto
συντελεστής συσχέτισης Pearson	coeficiente de correlação de Pearson	коэффициент корреляции Пирсона	coeficiente de correlación de Pearson
πεντασφαιρικό ημιβαριόγραμμα	variograma penta-esférico	пентасферическая семивариограмма	semivariograma penta-esférico
εκατοστημόριο	percentil	процентиль	percentil
κριτήρια επιτρεπτότητας	critérios de coerência	критерии допустимости	criterios de permisibilidad
επιτρεπτοί συντελεστές τάξης v	coeficientes autorizados de ordem v	допустимые коэффициенты порядка v	coeficientes autorizados de orden v
επιτρεπτοί συνδιασμοί τάξης v	combinações autorizadas de ordem v	допустимые комбинации порядка v	combinaciones autorizadas de orden v
άξων	pivot	разрешающий	pivote

English	Chinese	French	German
platykurtic	低峰态	platycurtique	flachgipflig
point kriging	点克里格法	krigeage ponctuel	Punktkrigen
Poisson distribution	泊松分布	loi de Poisson	Poisson-Verteilung
Poisson process	泊松过程	processus de Poisson	Poissonscher Prozeß
polygon of influence	影响多边形	polygone d'influence	Einflußpolygon
polynomial in several variables	多变量多项式	polynôme a plusieurs variables	mehrdimensionales Polynom
polynomial of degree *n*	n 次多项式	polynôme d'ordre *n*	Polynom *n*-ten Grades
polyscopic	多目标的	polyscopique	polyskopisch
population	总体	population	Gesamtheit
positive definite	正定	défini positif	positiv definit
posterior distribution	后验分布	loi conditionelle	a-posteriori Verteilung
power semivariogram model	指数半变差函数模型	variogramme puissance	Potenzen-Semivariogrammodell
precision	精度	précision	Exaktheit
prediction	预测	prédiction	Vorhersage
pre-posterior distribution	预后验分布	loi pre-postérieure	pre-posterior Verteilung
principal component	主分量	composante principale	Hauptkomponente
principal diagonal of a matrix	矩阵的主对角线	diagonale principale d'une matrice	Hauptdiagonale einer Matrix
prior distribution	先验分布	loi a priori	a-priori Verteilung
probabilistic regional semivariogram	概率区域半变差函数(图)	variogramme régional probabilisé	regionales Wahrscheinlichkeitssemivariogramm
probability	概率	probabilité	Wahrscheinlichkeit
probability density function	概率密度函数	fonction de densité	Wahrscheinlichkeitsdichte
probability distribution	概率分布	distribution de probabilité	Wahrscheinlichkeitsverteilung
probability function	概率函数	fonction de probabilité	Wahrscheinlichkeitsfunktion
probability interval	概率区间	intervalle de probabilité	Wahrscheinlichkeitsintervall
probability kriging	概率克里格法	krigeage de probabilité	Wahrscheinlichkeitskrigen
probability measure	概率测度	mesure de probabilité	Wahrscheinlichkeitsmaß
probability space	概率空间	espace de probabilité	Wahrscheinlichkeitsraum
product of a scalar and a matrix	数与矩阵的积	produit d'une matrice par un scalaire	Produkt aus einem Skalar und einer Matrix
product of matrices	矩阵的积	produit de matrices	Matrizenprodukt
projection theorem	射影定理	théorème de projection	Projektionssatz
proportional effect	比例效应	effet proportionnel	Proportionalitätseffekt
proximal polygon	邻近多边形	polygone de Voronoi	nächstes Polygon
punctual kriging	点克里格法	krigeage ponctuel	Punktkrigen
pure nugget effect	纯块金效应	effet de pépite pur	reiner Nuggeteffekt
pure nugget effect semivariogram model	纯块金效应半变差函数模型	modèle d'effet de pépite	Semivariogrammodell des reinen Nuggeteffekts
Pythagorean theorem	勾股定理;毕达哥拉斯定理	théorème de Pythagore	Satz des Pythagoras
q-q plot	Q-Q 图	q-q graphe	Quantildiagramm
quantile	分位数	quantile	Quantil
quartile	四分位数	quartile	Quartil
quartile deviation	四分位偏差	déviation quartile	quartile Abweichung
quasi-stationary	准平稳的	quasi-stationnaire	Quasistationarität

Greek	Portuguese	Russian	Spanish
πλατοκυρτικός	platicúrtica	плосковершинное	platicúrtico, achatado
σημειακό κρίκιγκ	krigagem pontual	точечный крайгинг	krigeage puntual
κατανομή Poisson	distribuição de Poisson	распределение Пуассона	distribución de Poisson
διαδικασία Poisson	processo de Poisson	Пуассоновский процесс	proceso de Poisson
πολύγωνο επιρροής	polígono de influência	полигон влияния	polígono de influencia
πολυώνυμο διαφόρων μεταβλητών	polinómio em várias variáveis	многочлен от многих переменных	polinomio en varias variables
πολυώνυμο n βαθμού	polinómio de grau n	многочлен степени н	polinomio de grado n
πολυσκοπικός	poliscópico	полископический	poliscópico
πληθυσμός	população	популяция, совокупность	población
γνησίως θετική	definido positivo	положительный определенный	definido positivo
μεταγενέστερη κατανομή	distribuição condicionada	апостериорное распределение	distribución a posteriori
ημιβαριόγραμμα δύναμης, δυναμοημιβαριόγραμμα	modelo de variograma com expoente	степенная модель семивариограммы	modelo de semivariograma potencial
επακρίβεια	precisão	точность	precisión
πρόβλεψη	previsão	прогноз	predicción
προμεταγενέστερη κατανομή	distribuição pré-posterior	преапостериорное распределение	distribución pre-posterior
κύρια συνιστώσα	componente principal	главная компонента	componente principal
κύρια διαγώνιος ενός πίνακα	diagonal principal de uma matriz	главная диагональ матрицы	diagonal principal de una matriz
προτέρα κατανομή	distribuição à priori	априорное распределение	distribución a priori
πιθανοτικό χωρικό ημιβαριόγραμμα	variograma regional probabilizado	вероятностная региональная семивариограмма	semivariograma probabilístico regional
πιθανότης	probabilidade	вероятность	probabilidad
συνάρτηση πυκνότητος πιθανότητος	função densidade de probabilidade	функция плотности вероятности	función de densidad de probabilidad
κατανομή πιθανότητας	distribuição de probabilidade	распределение вероятности	distribución de probabilidad
συνάρτηση πιθανότητας	função de probabilidade	функция вероятности	función de probabilidad
διάστημα πιθανότητας	intervalo de probabilidade	интервал вероятности	intervalo de probabilidad
πιθανοτικό κρίκιγκ	krigagem de probabilidade	крайгинг вероятности	krigeage de probabilidad
μέτρο πιθανότητας	medida de probabilidade	мера вероятности	medida de probabilidad
χώρος πιθανότητας	espaço de probabilidade	пространство вероятности	espacio probabilístico
γινόμενο αριθμού επί πίνακα	produto de um escalar por uma matriz	произведение матрицы на число	producto de un escalar por una matriz
γινόμενο πινάκων	produto de matrizes	произведение матриц	producto matricial
προβολικό θεώρημα	teorema de projecção	теорема проекции	teorema de la proyección
αναλογική επίδραση	efeito proporcional	пропорциональный эффект	efecto de proporcionalidad
εγγύς πολύγωνο	polígono de Voronoi	близкий полигон	polígono de proximidad
σημειακό κρίκιγκ	krigagem pontual	точечный крайгинг	krigeage puntual
απλή επίδραση κόκκου	efeito de pepita	чистый эффект самородков	efecto de pepita puro
ημιβαριόγραμμα απλής επίδρασης κόκκου	variograma pepítico	модель семивариограммы чистого эффекта самородков	semivariograma de efecto de pepita puro
Πυθαγόρειο θεώρημα	teorema de Pitágoras	теорема Пифагора	teorema de Pitágoras
σχέδιο ποσοστημόριου	diagrama q-q	диаграмма квантиль	diagrama de cuantiles
ποσοστημόριο	quantil	квантиль	cuantil
τεταρτημόριο	quartil	квартиль	cuartil
απόκλιση τεταρτημορίου	intervalo inter-quartílico	квартильное отклонение	desviación cuartílica
σχεδόν-στάσιμος	quase-estacionário	квазистационарный	cuasi-estacionario

English	Chinese	French	German
random coin method	随机掷硬币法	méthode des jetons aléatoires	Methode der zufälligen Auszählfläche
random experiment	随机试验	expérience aléatoire	Zufallsversuch
random field	随机场	champ aléatoire	Zufallsfeld
random function	随机函数	fonction aléatoire	Zufallsfunktion
random kriging	随机克里格法	krigeage aléatoire	Zufallskrigen
random process	随机过程	processus aléatoire	Zufallsprozeß
random set	随机集	ensemble aléatoire	Zufallsmenge
random tokens method	随机记号法	méthode des jetons aléatoires	Methode der zufälligen Auszählfläche
random variable	随机变量	variable aléatoire	Zufallsvariable
range	变程	a) étendue, b) portée	Reichweite
rank order transform	秩序变换	transformée de rang d'ordre	Rangordnungstransformierte
rank order variogram	秩序变差函数(图)	variogramme de rang d'ordre	Rangordnungsvariogramm
rational expression	有理表达式	expression rationelle	rationaler Ausdruck
realization	实现	réalisation	Realisierung
recoverable reserves	可回采储量	réserves récupérables	gewinnbare Reserven
recovery function	回采函数	fonction de récupération	Ausbringungsfunktion
recursion	递归	récursivité	Rekursion
reduced residual	约化剩余	résidu standard	reduziertes Residuum
regional characteristic	区域特征量	grandeur régionale	regionaler Kennwert
regional characteristic principle	区域特征量原则	principe des grandeurs régionales	Prinzip des regionalen Kennwertes
regionalized variable	区域化变量	variable régionalisée	ortsabhängige Variable
regional magnitude	区域量	grandeur régionale	regionale Größe
regional magnitude principle	区域量原则	principe des grandeurs régionales	Prinzip der regionalen Größen
regional semivariogram	区域半变差函数(图)	variogramme régional	regionales Semivariogramm
regression	回归	régression	Regression
regression line	回归线	ligne de régression	Regressionsgerade
regularization	正则化	régularisation	Vergleichmäßigung
regularized semivariogram	正则化半变差函数(图)	variogramme régularisé	vergleichmäßigtes Semivariogramm
regular system	正则系统	système régulier	reguläres System
relative nugget effect	相对块金效应	effet de pépite relatif	relativer Nuggeteffekt
relative semivariogram	相对半变差函数(图)	variogramme relatif	relatives Semivariogramm
representation of an intrinsic random function	内蕴随机函数的表示法	représentation d'une fonction aléatoire intrinsèque	Darstellung einer intrinsischen Zufallsfunktion
reserves in place	原地储量	réserves in situ	in-situ Vorräte, globale Vorräte
residual	剩余	résidu	Restgröße
residual kriging	剩余克里格法	krigeage des résidus	Krigen der Residuen
resistant	稳定的	resistant	unempfindlich
robust	稳健的	robustesse	robust
rodogram	方根变差函数	rodogramme	Rodogramm
Rodrigues formula	罗德里格斯公式	formule de Rodrigues	Rodriguessche Formel

Greek	Portuguese	Russian	Spanish
μέθοδος τυχαίου κέρματος	método dos círculos aleatórios	метод случайного выигрыша поверхности	método de las monedas aleatorias
τυχαίο πείραμα	experiência aleatória	случайный эксперимент	experimento aleatorio
τυχαίο πεδίο	campo aleatório	случайное поле	campo aleatorio
τυχαία συνάρτηση	função aleatória	случайная функция	función aleatoria
τυχαίο κρίκιγκ	krigagem aleatória	случайный крайгинг	krigeage aleatorio
τυχαία διαδικασία	processo aleatório	случайный процесс	proceso aleatorio
τυχαίο σύνολο	conjunto aleatório	случайное множество	conjunto aleatorio
μέθοδος τυχαίου κέρματος	método dos círculos aleatórios	метод случайного выигрыша поверхности	método de las monedas aleatorias
τυχαία μεταβλητή	variável aleatória	случайная переменная	variable aleatoria
εμβέλεια, κύμανση	alcance, amplitude	интервал влияния	a) amplitud, b) alcance
μετασχηματισμός βαθμού διάταξης	transformação por ordenação crescente	преобразование рангого порядка	transformada de orden secuencial
βαριόγραμμα βαθμού διάταξης	variograma da transformada por ordenação crescente	вариограмма рангого порядка	variograma de orden secuencial
λογική έκφραση	expressão racional	рациональное выражение	expresión racional
δειγματοσυνάρτηση	realização	реализация	realización
ανακτώμενα αποθέματα	reservas recuperáveis	балансовые запасы	reservas recuperables
συνάρτηση ανάκτησης	função de recuperação	функция извлечения	función de recuperación
αναγωγικός	recursividade	рекурсия	recursividad
περιορισμένο υπόλοιπο	resíduo reduzido	приведенный остаток	residuo reducido
τοπικό γνώρισμα	característica regional	региональная характеристика	característica regional
αρχή του τοπικού γνωρίσματος	princípio da característica regional	принцип региональной характеристики	principio de la característica regional
χωρική μεταβλητή	variável regionalizada	локализованная переменная	variable regionalizada
χωρικό μέγεθος	grandeza regional	региональная величина	característica regional
νόμος χωρικού μεγέθους	princípio das grandezas regionais	принцип региональной величины	principio de la característica regional
χωρικό ημιβαριόγραμμα	variograma regional	региональная семивариограмма	semivariograma regional
παλινδρόμηση	regressão	регрессия	regresión
γραμμή παλινδρόμησης	recta de regressão	линия регрессии	recta de regresión
ομαλοποίηση	regularização	регуляризация	regularización
ομαλοποιημένο ημιβαριόγραμμα	variograma regularizado	регуляризованная семивариограмма	semivariograma regularizado
κανονικό σύστημα	sistema determinado	регулярная система	sistema regular
σχετική επίδραση κόκκου	efeito de pepita relativo	относительный эффект самородков	efecto de pepita relativo
σχετικό ημιβαριόγραμμα	variograma relativo	относительная семивариограмма	semivariograma relativo
αναπαράσταση μιάς εγγενούς τυχαίας συνάρτησης	representação de uma função aleatória intrínsica	представление внутренней случайной функции	representación de una función aleatoria intrínseca
επιτόπου αποθέματα	reservas in situ	глобальные запасы	reservas in situ
υπόλοιπο	resíduo	остаток	residuo
κρίκιγκ υπολοίπου	krigagem dos resíduos	остаточный крайгинг	krigeage residual
ανθεκτικός	resistente	сопротивляющийся	resistente
εύρωστος	robustez	устойчивый, робастный	robusto
ροδόγραμμα	rodograma	родограмма	rodograma
τύπος του Rodrigues	fórmula de Rodrigues	формула Родригесса	fórmula de Rodrigues

English	Chinese	French	German
rose diagram	玫瑰图	graphe en coordonées polaires	Rose-Diagramm
row of a determinant	行列式的行	ligne d'un déterminant	Zeile in einer Determinante
row of a matrix	矩阵的行	ligne d'une matrice	Zeile einer Matrix
σ-algebra	σ-代数	σ-algèbre	σ-Algebra
sample	样本;样品	échantillon	a) Stichprobe, b) Probe
sample description space	样本描述空间	espace des états d'un système	Stichprobenraum
sample function	样本函数	fonction d'échantillonnage	Stichprobenfunktion
sample size	样本容量	taille de l'énchantillon	Stichprobenumfang
sample space	样本空间	espace des échantillons	Stichprobenraum
scalar product	标积	produit scalaire	Skalarprodukt
scattergram	散点图	nuage de corrélation	Streudiagramm
scatterplot	散点图	graphe de dispersion	Streudiagramm
Schwarz inequality	许瓦尔兹不等式	inégalité de Schwarz	Schwarzsche Ungleichung
screen effect	屏蔽效应	effet d'ecran	Abschirmeffekt
second order random field	二阶随机场	champ aléatoire d'ordre 2	Zufallsfeld 2. Ordnung
second order random variable	二阶随机变量	variable aléatoire d'ordre 2	Zufallsvariable 2. Ordnung
selective	选择性的	sélectif	selektiv
selectivity curve	选择性曲线	courbe de sélectivité	Selektivitätskurve
selfsimilar	自相似	auto-homothétique	selbstähnlich
semimarkovian random function	半马尔可夫随机函数	fonction aléatoire semi-markovienne	semimarkovsche Zufallsfunktion
semivariogram	半变差函数(图)	demi-variogramme	Semivariogramm
semivariogram fitting	半变差函数拟合	ajustement du variogramme	Anpassung eines Variogramms
semivariogram model	半变差函数模型	variogramme modélisé	Semivariogrammodell
sequential indicator simulation	序列指标模拟	simulation sequentielle d'indicatrices	sequentielle Indikatorsimulation
serial variation function	序列变异函数;半变差函数	serial variation function	serielle Variationsfunktion
Sichel t distribution	西切尔 t 分布	loi de Sichel	Sichelsche t-Verteilung
sigma-field	σ- 域	sigma-algèbre	Sigmafeld
sill	基台	palier	Schwelle
similarity dimension	相似性维数	dimension de similitude	fraktale Dimension
simple kriging	简单克里格法	krigeage simple	einfaches Krigen
simulation	模拟	simulation	Simulation
single indicator kriging	单指标克里格法	krigeage d'indicatrice unique	einfaches Indikatorkrigen
singular matrix	奇异矩阵	matrice singulière	singuläre Matrix
skewness	偏度	disymétrie	Schiefe
sliding neighborhood	滑动邻域	voisinage glissant	gleitende Nachbarschaft
sliding representation	滑动表示法	représentation glissante	gleitende Darstellung
smoothing	平滑	lissage	Glättung
soft data	软数据	donnée molle	weiche Daten

Greek	Portuguese	Russian	Spanish
διάγραμμα ρόδου	diagrama polar	диаграмма розы	diagrama de roseta
γραμμή μιάς ορίζουσας	linha de um determinante	строка определителя	fila de un determinante
γραμμή ενός πίνακα	linha de uma matriz	строка матрицы	fila de una matriz
σ-άλγεβρα	álgebra-σ	σ-алгебра	álgebra σ
δείγμα	amostra	выборка	muestra
δειγματοχώρος	espaço dos resultados, espaço amostral	выборочное дескриптивное пространство	espacio muestral
δειγματοσυνάρτηση	função de amostragem	функция выборки, выборочная функция	función de muestreo
μέγεθος δείγματος	dimensão da amostra	объем выборки	tamaño de la muestra
χώρος δείγματος	espaço das amostras	пространство выборок, выборочная функция	espacio muestral
εσωτερικό γινόμενο	produto escalar	скалярное произведение	producto escalar
διασπορόγραμμα	núvem de correlação	поле рассеивания	diagrama de dispersión
διασπορόγραμμα	núvem de correlação	диаграмма рассеивания	diagrama de dispersión
ανισότης Schwarz	desigualdade de Schwarz	неравенство Шварца	desigualdad de Schwarz
επίδραση οθόνης	efeito de écran	эффект просеивания, эффект отсеивания	efecto pantalla
τυχαίο πεδίο δευτέρας τάξης	campo aleatório de ordem 2	случайное поле второго порядка	campo aleatorio de segundo orden
τυχαία συνάρτηση δευτέρας τάξης	variável aleatória de ordem 2	случайная переменная второго порядка	variable aleatoria de segundo orden
εκλεκτικός	selectiva	селективный	selectivo
καμπύλη εκλεκτικότητος	curva de selectividade	кривая селективности	curva de selectividad
αυτο-ομοθετικός	autohomotético	автоподобный	autohomotético, autosemejante
τυχαία συνάρτηση ημι-Markov	função aleatória semimarkoviana	полумарковская случайная функция	función aleatoria semimarkoviana
ημιβαριόγραμμα	semivariograma	семивариограмма	semivariograma
προσαρμογή ημιβαριογράμματος	ajustamento do variograma	аппроксимация семивариограммы	ajuste de un semivariograma
μοντέλο ημιβαριογράμματος	modelo de variograma	модель семивариограммы	modelo de un semivariograma
ακολουθιακή προσομοίωση δείκτη	simulação sequencial da indicatriz	моделирование последовательного индикатора	simulación secuencial de un indicador
συνεχόμενη συνάρτηση μεταβολής	semivariograma (segundo Jowett & Matérn)	функция сериалной дисперсии	función de variación en serie
Sichel t κατανομή	distribuição t de Sichel	т-распределение Сихеля	distribución t de Sichel
πεδίο σίγμα	álgebra-σ	сигма-поле	campo sigma
κατώφλι	patamar	порог, силь	meseta
διάσταση ομοιότητος	dimensão de similaridade	размер подобия, размерность подобия	dimensión de similitud
απλό κρίκιγκ	krigagem simples	простой крайгинг	krigeage simple
προσομοίωση	simulação	моделирование, имитация	simulación
μοναδικό κρίκιγκ δείκτη	krigagem simples da indicatriz	единичный индикаторный крайгинг	krigeage de un solo indicador
ιδιάζων πίνακας	matriz singular	особенная матрица	matriz singular
λοξότης	assimetria	асимметрия	asimetría
ολισθαίνουσα γειτονιά	vizinhança deslizante	скользящая окрестность	entorno móvil
ολισθαίνουσα αναπαράσταση	representação deslizante	скользящее представление	representación móvil
εξομάλυνση	atenuação, regularização	сглаживание	suavizamiento
μαλακά δεδομένα	dado vago, dado soft	неточные данные	información inexacta

English	Chinese	French	German
space transformation of the first kind	第一类空间变换	transformation spatiale de classe 1	Raumtransformation 1. Art
space transformation of the second kind	第二类空间变换	transformation spatiale de classe 2	Raumtransformation 2. Art
spatial dependence	空间相关	corrélation spatiale	räumliche Abhängigkeit
spatial independence	空间无关	non-corrélation spatiale	räumliche Unabhängigkeit
Spearman correlation coefficient	斯皮尔曼相关系数	coefficient de correlation de Spearman	Spearmanscher Korrelationskoeffizient
spherical semivariogram	球状半变差函数(图)	variogramme sphérique	sphärisches Semivariogramm
spline function	样条函数	fonction spline	Spline
square matrix	方阵	matrice carrée	quadratische Matrix
standard deviation	标准离差	ecart-type	Standardabweichung
standard error	标准误差	erreur-type	Standardfehler
standard normal distribution	标准正态分布	loi normale	standardisierte Normalverteilung
standard score	标准得分	note réduit	standardisierte Punktwerte
stationarity	平稳性	stationnarité	Stationarität
stationary	平稳的	stationnaire	stationär
stationary in the wide sense	广义平稳的	stationnaire au sens large	stationär im weiteren Sinne
stationary of finite order	有限阶平稳的	stationnarité d'ordre fini	stationär endlicher Ordnung
statistic	统计量	statistique	Statistik
statistical dependence	统计相关	dépendance statistique	statistische Abhängigkeit
statistical independence	统计无关	indépendance statistique	statistische Unabhängigkeit
step function	阶梯函数	fonction échelon	Stufenfunktion
stochastic dependence	随机相关	dépendance stochastique	stochastische Abhängigkeit
stochastic independence	随机无关	indépendance stochastique	stochastische Unabhängigkeit
stochastic process	随机过程	processus stochastique	stochastischer Prozeß
stochastic variable	随机变量	variable stochastique	stochastische Variable
stratified anisotropy	分层各向异性	anisotropie zonale	zonale Anisotropie
stratified random sampling	分层随机取样	échantillonnage aléatoire stratifié	geschichtet zufällige Probennahme
strictly conditionally positive definite function	严格条件正定函数	fonction strictement définie positive conditionnelle	streng bedingt positiv definite Funktion
strictly positive definite function	严格正定函数	fonction strictement définie positive	streng positiv definite Funktion
strict stationarity	严格平稳性	stationnarité stricte	strenge Stationarität
strong stationarity	强平稳性	stationnarité stricte	strenge Stationarität
structural analysis	结构分析	analyse structurale	Strukturanalyse
structural function	结构函数	fonction structurale	strukturelle Funktion
structure	结构	structure	Struktur
structure function	结构函数	fonction de structure	Strukturfunktion
support	承载	support	a) Stützung einer regionalisierbaren Variablen, b) Träger einer Funktion
support effect	承载效应	effet de support	Stützungseffekt

Greek	Portuguese	Russian	Spanish
χωρικός μετασχηματισμός πρώτου είδους	transformação de primeira espécie	преобразование пространства первой степени	transformación espacial de primera especie
χωρικός μετασχηματισμός δευτέρου είδους	transformação de segunda espécie	преобразование пространства второй степени	transformación espacial de segunda especie
χωρική εξάρτηση	dependência espacial	пространственная зависимость	dependencia espacial
χωρική ανεξαρτησία	independência espacial	пространственная независимость	independencia espacial
συντελεστής συσχέτισης Spearman	coeficiente de correlação de Spearman	коэффициент корреляции Спирмана	coeficiente de correlación de Spearman
σφαιρικό ημιβαριόγραμμα	variograma esférico	сферическая семивариограмма	semivariograma esférico
συνάρτηση σπλίνας	função spline	сплайн-функция	spline
τετραγωνικός πίνακας	matriz quadrada	квадратная матрица	matriz cuadrada
τυπική απόκλιση	desvio-padrão	стандартное отклонение	desviación estándard
τυπικό λάθος	erro padrão	стандартная ошибка	error estándard
τυπική κανονική κατανομή	distribuição normal reduzida, distribuição normal normalizada	стандартное нормальное распределение	distribución normal estandardizada, distribución normal estandarizada
τυπικός βαθμός	contagem normalizada	стандартное z-значение	puntaje estándar
στασιμότης	estacionariedade	стационарность	estacionariedad
στάσιμος	estacionário	стационарный	estacionario
στάσιμος με την ευρεία έννοια	estacionário em sentido lato	стационарный в широком смысле	estacionario en sentido amplio
στάσιμος πεπερασμένης τάξης	estacionário de ordem finita	стационарный конечного порядка	estacionario de orden finito
στατιστική	estatístico	статистика	estadístico
στατιστική εξάρτηση	dependência estatística	статистическая зависимость	dependencia estadística
στατιστική ανεξαρτησία	independência estatística	статистическая независимость	independencia estadística
βηματική συνάρτηση	função em escada	ступенчатая функция	función escalón
στοχαστική εξάρτηση	dependência estocástica	стохастическая зависимость	dependencia estocástica
στοχαστική ανεξαρτησία	independência estocástica	стохастическая независимость	independencia estocástica
στοχαστική διαδικασία	processo estocástico	стохастический процесс	proceso estocástico
στοχαστική μεταβλητή	variável estocástica	стохастическая переменная	variable estocástica
στρωματοποιημένη ανισοτροπία	anisotropia zonal	стратифицированная анизотропия	anisotropía estratificada
στρωματοποιημένη τυχαία δειγματοληψία	amostragem aleatória estratificada	расслоенный случайный выбор	muestreo aleatorio estratificado
αυστηρά δεσμευμένη γνησίως θετική συνάρτηση	função estritamente definida positiva condicional	строго условно положительная определенная функция	función estrictamente definida condicionalmente positiva
αυστηρά γνησίως θετική συνάρτηση	função estritamente definida positiva	строго положительная определенная функция	función estrictamente definida positiva
αυστηρή στασιμότης	estacionariedade estrita	точная стационарность	estacionariedad estricta
ισχυρή στασιμότης	estacionariedade forte	сильная стационарность	estacionariedad estricta
δομική ανάλυση	análise estrutural	структурный анализ	análisis estructural
δομική συνάρτηση	função estrutural	структурная функция	función estructural
δομή, κατασκευή	estrutura	структура	estructura
δομική συνάρτηση	função estrutural	функция структуры	función de estructura
υποστήριξη, φορέας	suporte	суппорт, носитель локализованной переменной	soporte
επίδραση φορέα	efeito de suporte	эффект носителя множества	efecto soporte

English	Chinese	French	German
symmetric matrix	对称矩阵	matrice symétrique	symmetrische Matrix
theoretical semivariogram	理论半变差函数(图)	variogramme théorique	theoretisches Semivariogramm
theory	理论	théorie	Theorie
Thiessen polygon	替森多边形	polygone de Thiessen	Thiessen-Polygon
three-parameter lognormal distribution	三参数对数正态分布	loi lognormale à trois paramètres	dreiparametrische lognormale Verteilung
time series	时间序列	série temporelle	Zeitreihe
topoprobabilistic model	地形概率模型	modèle topo-probabiliste	topologisches Wahrscheinlichkeitsmodell
transfer function	转换函数	fonction de transfert	Transferfunktion
transform	变换	transformée	Transformierte
transformation	变换	transformation	Transformation
transition probability	转移概率	probabilité de transition	Übergangswahrscheinlichkeit
transitive	跃迁的	transitif	transitiv
transitive covariogram	跃迁协变差函数	covariogramme transitif	transitives Kovariogram
transitive kriging	跃迁克里格法	krigeage transitif	transitives Krigen
transitive method	跃迁法	méthode transitive	transitive Methoden
translation of permissible coefficients	允许系数的平移	translation de coefficients autorisés	Translation zulässiger Koeffizienten
transpose of a matrix	矩阵的转置	matrice transposée	Transponierte einer Matrix
trend	趋势	tendance régionale	Trend
trial	试验	tirage	Versuch
turning bands method	旋转带法	méthode des bandes tournantes	Methode der drehenden Bänder
two-parameter lognormal distribution	二参数对数正态分布	loi lognormale à deux paramètres	zweiparametrische lognormale Verteilung
unbiased	无偏的	sans biais	unverzerrt
uncertainty	不确定性	incertitude	Ungewißheit
underestimation	过低估计	sous-estimation	Unterschätzung
underlying semivariogram	基础半变差函数(图)	variogramme sous-jacent	zugrundeliegendes Semivariogramm
uniform conditioning	一致条件化	conditionnement uniforme	gleichmäßiges Bedingen
uniform distribution	均匀分布	loi uniforme	gleichmäßige Verteilung, Gleichverteilung
uniform transformation	均匀变换	transformée uniforme	Transformation in eine Gleichverteilung
unimodal distribution	单峰分布	distribution monomodale	unimodale Verteilung
union stability	并集稳定性	stabilité pour la réunion	Vereinigungsstabilität
univariate	一元的	univariable	eindimensional
universal	泛的	universel	universell
universality condition	泛性条件	condition d'universalité	Universalitätsbedingung
universal kriging	泛克里格法	krigeage universel	Universalkrigen
universe	整体;全域	population	Gesamtheit
upper quartile	上四分位数	quartile supérieur	oberes Quartil
validation	确认	validation	Gültigkeitstest
variance	方差	variance	Varianz

Greek	Portuguese	Russian	Spanish
συμμετρικός πίνακας	matriz simétrica	симметрическая матрица	matriz simétrica
θεωρητικό ημιβαριόγραμμα	variograma teórico	теоретическая семивариограмма	semivariograma teórico
θεωρία	teoria	теория	teoría
πολύγωνο Thiessen	polígono de Thiessen	полигон Тиссена	polígono de Thiessen
λογοκανονική κατανομή τριών παραμέτρων	distribuição lognormal a três parâmetros	трехпараметрическое логнормальное распределение	distribución lognormal de tres parámetros
χρονοσειρά	série temporal	временный ряд	serie temporal
τοποπιθανοτικό μοντέλο	modelo topoprobabilístico	топовероятностная модель	modelo topoprobabilístico
συνάρτηση μεταφοράς	função de transferência	передаточная функция	función de transferencia
μετασχηματίζω	transformada	преобразование	transformada
μετασχηματισμός	transformação	преобразование	transformación
μεταβατική πιθανότης	probabilidade de transição	вероятность перехода	probabilidad de transición
μεταβατικός	transitivo	транзитивный	transitivo
μεταβατικό συνδιασπορόγραμμα	covariograma transitivo	транзитивная ковариограмма	covariograma transitivo
μεταβατικό κρίκιγκ	krigagem transitiva	транзитивный крайгинг	krigeage transitivo
μεταβατικές μέθοδοι	método transitivo	транзитивный метод	método transitivo
μεταφορά επιτρεπτών συντελεστών	translacção de coeficientes autorizados	трансляция допустимых коэффициентов	translación de coeficientes autorizados
ανάστροφος ενός πίνακα	matriz transposta	транспонировать матрицу	matriz transpuesta
τάση	tendência	направление, тренд	tendencia
δοκιμή	ensaio	случайный эксперимент	intento, ensayo, prueba
μέθοδος στρεφόμενης δέσμης	método das bandas rotativas	метод вращающихся полосок	método de las bandas rotantes
λογοκανονική κατανομή δύο παραμέτρων	distribuição lognormal a dois parâmetros	двухпараметрическое логнормальное распределение	distribución lognormal de dos parámetros
αμερόληπτος	não enviesado	несмещенный	insesgado
αβεβαιότης	incerteza	неопределенность	incertidumbre
υποεκτίμηση	subestimação	недооценка, недооценивание	subestimación
ενυπάρχον ημιβαριόγραμμα	variograma subjacente	семивариограмма для остатков локализованных переменных	semivariograma subyacente
ομοιόμορφη δέσμευση	condicionamento uniforme	равномерное кондиционирование	condicionamiento uniforme
ομοιόμορφη κατανομή	distribuição uniforme	равномерное распределение	distribución uniforme
ομοιόμορφος μετασχηματισμός	transformação uniforme	равномерное преобразование	transformación uniforme
μονόκορφη κατανομή	distribuição unimodal	унимодальное распределение	distribución unimodal
ευστάθεια ένωσης	estabilidade para a união	устойчивость объединения	estabilidad de una unión
μονομεταβλητή	unidimensional	одномерная переменная	univariable
καθολικός	universal	универсальный	universal
καθολική συνθήκη	condição de universalidade	универсальное условие	restricción universal
καθολικό κρίκιγκ	krigagem universal	универсальный крайгинг	krigeage universal
πληθυσμός, ολότητα	universo, população	генеральный	universo
άνω τεταρτημόριο	terceiro quartil	верхний квартиль	cuartil superior
επικύρωση	validação	обоснованность	validación
διασπορά	variância	дисперсия	varianza

English	Chinese	French	German
variance adjustment factor	方差调整因子	facteur de correction de variance	Varianzkorrekturfaktor
variate	变量	variable aléatoire	Variable
variogram	变差函数(图)	variogramme	Variogramm
variogram of order α	α阶变差函数(图)	variogramme d'ordre α	Variogramm der Ordnung α
variography	变差函数分析	variographie	Variographie
vector	向量	vecteur	Vektor
vector space	向量空间	espace vectoriel	Vektorraum
Voronoi division	沃罗诺瓦剖分	partition de Voronoi	Voronoi-Zerlegung
Voronoi polygon	沃罗诺瓦多边形	polygone de Voronoi	Voronoi-Polygon
Voronoi tesselation	沃罗诺瓦嵌形	mosaique de Voronoi	Voronoi-Parkettierung
weak stationarity	弱平稳性	stationnarité faible	schwache Stationarität
white noise	白噪声	bruit blanc	weißes Rauschen
white random function	白随机函数	fonction aléatoire en bruit blanc	Zufallsfunktion des weißen Rauschens
Wiener filter	维纳滤波	filtre de Wiener	Wiener–Filter
Wiener–Levy process	维纳一勒维过程	processus de Wiener-Lévy	Wiener–Levy-Prozeß
Wiener process	维纳过程	processus de Wiener	Wiener-Prozß
Wigner–Seitz cell	维格纳-塞茨格子	cellule de Wigner–Seitz	Wigner–Seitz-Zelle
window	窗	fenêtre	Fenster
working neighborhood	工作邻域	voisinage de travail	Arbeitsbereich
working scale	工作比例尺	échelle de travail	Arbeitsmaßstab
zonal anisotropy	带状各向异性	anisotropie zonale	zonale Anisotropie
zone of influence	影响带	zone d'influence	Einflußbereich
z-score	Z 得分	cote centré reduit	standardisierte Punktwerte

Greek	Portuguese	Russian	Spanish
παράγων προσαρμογής διασποράς	factor de ajustamento da variância	дисперсионный коэффициент выравнивания	factor de ajuste de la varianza
τυχαία μεταβλητή	variável aleatória	случайная величина	variable aleatoria
βαριόγραμμα	variograma	вариограмма	variograma
βαριόγραμμα τάξης α	variograma de ordem α	вариограмма порядка α	variograma de orden α
βαριογραφία	variografia	вариография	variografía
διάνυσμα	vector	вектор	vector
διανυσματικός χώρος	espaço vectorial	векторное пространство	espacio vectorial
διαίρεση Voronoi	divisão de Voronoi	разделение Вороного	división de Voronoi
πολύγωνο Voronoi	polígono de Voronoi	полигон Вороного	polígono de Voronoi
μωσαϊκό Voronoi	tecelagem de Voronoi	замощение плоскости Вороного	teselado de Voronoi
ασθενής στασιμότης	estacionariedade fraca	слабая стационарность	estacionariedad débil
λευκός θόρυβος	ruído branco	белый шум	ruido blanco
λευκή τυχαία συνάρτηση	função aleatória de ruído branco	случайная функция белого процесса	función aleatoria de ruido blanco
φίλτρο Wiener	filtro de Wiener	фильтр Винера	filtro de Wiener
διαδικασία Wiener–Levy	processo de Wiener–Levy	процесс Винера-Леви	proceso de Wiener–Levy
διαδικασία Wiener	processo de Wiener	процесс Винера	proceso de Wiener
κελί Wigner–Seitz	célula de Wigner–Seitz	клетка, ячейка Вигнера-Зайтца	celda de Wigner–Seitz
παράθυρο	janela	окно, часть поля	ventana
γειτονιά εργασίας	vizinhança de trabalho	рабочая окрестность	entorno de trabajo
κλίμακα εργασίας	escala de trabalho	рабочий масштаб	escala de trabajo
ζωνική ανισοτροπία	anisotropia zonal	зональная анизотропия	anisotropía zonal
ζώνη επιρροής	zona de influência	зона влияния	zona de influencia
αποτέλεσμα z	contagem normalizada	з-"score", з-значение	puntaje z

CHINESE to ENGLISH

E 型估计值 E-type estimate
F 函数 F-function
h 散点图 h-scattergram
LU 分解法 lower-upper (LU) decomposition
***n* 次多项式** polynomial of degree *n*
Q-Q 图 q-q plot
Z 得分 z-score
α 阶变差函数(图) variogram of order α
β 分布 beta distribution
β 函数 beta function
γ 分布 gamma distribution
γ 函数 gamma function
σ 域 sigma-field
ν 阶广义协方差 generalized covariance of order ν
ν 阶广义增量 generalized increments of order ν
ν 阶内蕴随机函数 intrinsic random function of order ν
ν 阶条件正定 conditionally positive definite of order ν
ν 阶允许系数 authorized coefficients of order ν; permissible coefficients of order ν
ν 阶允许组合 authorized combinations of order ν; permissible combinations of order ν
σ–代数 σ-algebra
埃尔米特多项式 Hermite polynomials
埃尔米特矩阵 Hermitian matrix
埃尔米特模型 Hermite model
白随机函数 white random function
白噪声 white noise
百分位数 centile; percentile
半变差函数 serial variation function
半变差函数(图) semivariogram; demivariogram
半变差函数模型 semivariogram model
半变差函数拟合 semivariogram fitting
半马尔可夫随机函数 semimarkovian random function
贝努里试验 Bernoulli experiment
贝塞尔不等式 Bessel inequality
贝叶斯定理 Bayes' theorem
比例效应 proportional effect
毕达哥拉斯定理 Pythagorean theorem
闭合的 closed
边缘分布 marginal distribution
边缘概率密度函数 marginal probability density function
变差函数(图) variogram
变差函数分析 variography
变程 range
变换 transform; transformation
变量 variate
变量的线性组合 linear combination of variables
变形 anamorphosis
变异系数 coefficient of variation
遍历定理 ergodic theorem
遍历性 ergodicity
遍历性质 ergodic property
标积 scalar product
标准得分 standard score
标准离差 standard deviation
标准误差 standard error
标准正交函数 orthonormal function
标准正交向量 orthonormal vector
标准正态分布 standard normal distribution
并集的无限可分性 infinite divisibility of a union
并集稳定性 union stability
波莱尔域 Borel field
泊松分布 Poisson distribution
泊松过程 Poisson process
不确定性 uncertainty
布尔随机函数 Boolean random function
布朗运动过程 Brownian motion process
布尼亚科夫斯基不等式 Buniakovski inequality
操作概念 operational concept
测度 measure
场 field
常峰态 mesokurtic
超出系数 coefficient of excess
承载 support
承载变化模型 change of support model
承载效应 support effect
窗 window
纯块金效应 pure nugget effect
纯块金效应半变差函数模型 pure nugget effect semivariogram model
大数定律 law of large numbers
带状各向异性 zonal anisotropy
单峰分布 unimodal distribution
单目标的 monoscopic
单位矩阵 identity matrix
单指标克里格法 single indicator kriging
淡化随机函数 dilution random function
德劳内剖分 Delaunay subdivision
德劳内三角剖分 Delaunay triangulation
德威依斯半变差函数(图) de Wijsian semivariogram
德威依斯公式 de Wijs formula
等因子二元密度函数 isofactorial bivariate density function
等值线 contour; isoline; isopleth
低峰态 platykurtic
狄拉克 δ 函数 Dirac delta function
狄利克雷嵌形 Dirichlet tesselation
地形概率模型 topoprobabilistic model
地质统计学 geostatistics
第二类空间变换 space transformation of second kind

第一类空间变换 space transformation of first kind
递归 recursion
点克里格法 point kriging; punctual kriging
度量空间 metric space
对称矩阵 symmetric matrix
对角二元密度函数 diagonal bivariate density function
对角化矩阵 diagonalized matrix
对角矩阵 diagonal matrix
对偶克里格法 dual kriging
对数半变差函数(图) logarithmic semivariogram
对数克里格法 logarithmic kriging
对数正态分布 lognormal distribution
对数正态校正 lognormal correction
对数正态克里格法 lognormal kriging
对于第二个函数的可积函数 integrable function with respect to a second function
多变量多项式 polynomial in several variables
多重指标克里格法 multiple indicator kriging
多峰分布 multimodal distribution
多目标的 polyscopic
多元的 multivariate
多元分布 multivariate distribution
多元高斯克里格法 multigaussian kriging
多元正态分布 multinormal distribution
二参数对数正态分布 two-parameter lognormal distribution
二阶随机变量 second order random variable
二阶随机场 second order random field
二项分布 binomial distribution
二元的 bivariate
二元分布 bivariate distribution
二元高斯模型 bigaussian model
二元正态分布 bivariate normal distribution
反卷积 deconvolution
泛的 universal
泛克里格法 universal kriging
泛目标的 panscopic
泛性条件 universality condition
方差 variance
方差调整因子 variance adjustment factor
方根变差函数 rodogram
方阵 square matrix
方阵的阶 order of a square matrix
仿射变换 affinity transformation
仿射校正 affine correction
非遍历实验协方差 nonergodic experimental covariance
非参数地质统计学 nonparametric geostatistics
非奇异系统 nonsingular system
非线性地质统计学 nonlinear geostatistics
非中心协方差 noncentered covariance
分布 distribution
分布的矩 moment of a distribution
分布函数 distribution function
分布守恒假设 conservation of law hypothesis
分层各向异性 stratified anisotropy
分层随机取样 stratified random sampling
分数维 fractal dimension
分位数 fractile; quantile
分形 fractal
峰度 kurtosis
峰态系数 coefficient of kurtosis
负二项分布 negative binomial distribution
傅里叶变换 Fourier transform
傅里叶分析 Fourier analysis
傅里叶合成 Fourier synthesis
概率 probability
概率测度 probability measure
概率分布 probability distribution
概率函数 probability function
概率克里格法 probability kriging
概率空间 probability space
概率密度函数 probability density function
概率区间 probability interval
概率区域半变差函数(图) probabilistic regional semivariogram
高品位结构化 high-grade structuration
高品位去结构化 high-grade destructuration
高斯半变差函数(图) Gaussian semivariogram
高斯变形 Gaussian anamorphosis
高斯分布 Gaussian distribution
各向同性 isotropy
各向异性 anisotropy
共轭复数 conjugate complex number
工作比例尺 working scale
工作邻域 working neighborhood
勾股定理 Pythagorean theorem
构形 configuration
估计 estimation
估计方差 estimation variance
估计量 estimator
估计值 estimate
广义半变差函数(图) generalized semivariogram
广义平稳的 stationary in the wide sense
过低估计 underestimation
过高估计 overestimation
海维赛德函数 Heaviside's function
函数 function
行列式 determinant
行列式的行 row of a determinant
行列式的阶 order of a determinant
行列式的列 column of a determinant
行列式的元素 element of a determinant
行列式元素的余子式 cofactor of an element in a determinant
豪斯道夫-贝西科维奇维 Hausdorff–Besicovitch dimension
豪斯道夫测度 Hausdorff measure
后验分布 posterior distribution
互半变差函数(图) cross-semivariogram
互变差函数(图) cross-variogram
互相关 cross-correlation
互相关图 crosscorrelogram
互相关系数 cross-correlation coefficient

互协方差 cross-covariance
滑动表示法 sliding representation
滑动邻域 sliding neighborhood
回采函数 recovery function
回归 regression
回归线 regression line
基础半变差函数(图) underlying semivariogram
基函数 basis function
基台 sill
基准面 datum
积分范围 integral range
极大似然估计量 maximum likelihood estimator; maximum likelihood statistic
极大熵形式体系 maximum entropy formalism
极小均方误差估计量 minimum mean square error estimator
极小误差模估计量 minimum error norm estimator
级配 grading
级配半变差函数(图) graded semivariogram
集的分割 partition of a set
几何场 geometric field
几何承载 geometric support
几何各向异性 geometric anisotropy
几何平均值 geometric average; geometric mean
几何协变差函数 geometric covariogram
尖峰态 leptokurtic
间接对数校正 indirect lognormal correlation
简单克里格法 simple kriging
减弱丛聚 declustering
交叉验证 cross-validation
阶梯函数 step function
结构 structure
结构分析 structural analysis
结构函数 structural function; structure function
结果 outcome
紧的 compact
精度 precision
精确内插量 exact interpolator
局部的 local
局部估计 local estimation
局部相对半变差函数(图) local relative semivariogram
矩 moment
矩阵 matrix
矩阵的行 row of a matrix
矩阵的行列式 determinant of a matrix
矩阵的积 product of matrices
矩阵的列 column of a matrix
矩阵的逆 inverse of a matrix
矩阵的特征方程 characteristic equation of a matrix
矩阵的特征值 eigenvalue of a matrix
矩阵的元素 element of a matrix
矩阵的主对角线 main diagonal of a matrix; principal diagonal of a matrix
矩阵的转置 transpose of a matrix
聚点 accumulation point
卷积 convolution
绝对离散 absolute dispersion
均方误差 mean square error
均匀变换 uniform transformation
均匀分布 uniform distribution
均值 mean
卡蒂尔关系式 Cartier's relation
开的 open
柯西-许瓦尔兹不等式 Cauchy–Schwarz inequality
柯西序列 Cauchy sequence
可回采储量 recoverable reserves
可积函数 integrable function
可加集函数 additive set function
可加性定理 additivity theorem
可因子化随机过程 factorable random process
可因子化随机函数 factorable random function
客观涵义 objective meaning
客观性准则 objectivity criterion
克里格法 kriging
克里格方差 kriging variance
克里格公式 Krige's formula
克里格邻域 kriging neighborhood
克里格椭园 Krige's ellipse
克罗内克符号 Kronecker delta
空间无关 spatial independence
空间相关 spatial dependence
空穴效应 hole effect
空穴效应半变差函数模型 hole effect semivariogram model
块段克里格法 block kriging
块金效应 nugget effect
快速傅里叶变换 fast Fourier transform
拉盖尔多项式 Laguerre polynomials
拉格朗日乘子 Lagrange multiplier
拉格朗日乘子法 Lagrange's method of multipliers
拉普拉斯变换 Laplace transform
累积 accumulation
累积分布函数 cumulative distribution function
累积频率 cumulative frequency
离散方差 dispersion variance
离散高斯模型 discrete Gaussian model
离散随机变量 discrete random variable
理论 theory
理论半变差函数(图) theoretical semivariogram
立方半变差函数(图) cubic semivariogram
联合概率分布 joint probability distribution
联合概率密度函数 joint probability density function
联合交叉验证 joint cross-validation
联合可因子化随机函数 jointly factorable random function
联合正交随机过程 jointly orthogonal random process
联合正交随机函数 jointly orthogonal random function
连续型随机变量 continuous random variable
两两相对半变差函数(图) pairwise relative semivariogram
邻近多边形 proximal polygon
邻域 neighborhood
流形 manifold

滤波 filtering
罗德里格斯公式 Rodrigues formula
马尔可夫过程 Markov process
脉冲 impulse
玫瑰图 rose diagram
梅也伦格子 Meijering's cell
蒙特卡罗法 Monte Carlo method
模 norm
模糊指标数据 fuzzy indicator data
模拟 simulation
模型 model
目标函数 objective function
内部 interior
内插 interpolation
内积 inner product
内积空间 inner product space
内蕴的 intrinsic
内蕴函数 intrinsic function
内蕴假设 intrinsic hypothesis
内蕴克里格法 intrinsic kriging
内蕴离散 intrinsic dispersion
内蕴属性 intrinsic attribute
内蕴随机函数的表示法 representation of an intrinsic random function
内蕴相关 intrinsic correlation
逆问题 inverse problem
欧几里德空间 euclidean space
帕赛维尔方程 Parseval's equation
抛物线半变差函数(图) parabolic semivariogram
抛物线性态 parabolic behavior
皮尔逊相关系数 Pearson correlation coefficient
偏度 skewness
偏态系数 coefficient of skewness
偏倚 bias
漂移 drift
频率函数 frequency function
频率曲线 frequency curve
频率图 frequency diagram
屏蔽效应 screen effect
平滑 smoothing
平均 average
平均半变差函数(图) average semivariogram; mean semivariogram
平均值 average value; mean
平稳的 stationary
平稳性 stationarity
普通克里格法 ordinary kriging
期望值 expected value
齐性 homogeneity
奇异矩阵 singular matrix
强平稳性 strong stationarity
乔奎特容量 Choquet's capacity
乔勒斯基方法 Choleski's method
球状半变差函数(图) spherical semivariogram
趋势 trend
曲线拟合 curve fitting
区域半变差函数(图) regional semivariogram
区域化变量 regionalized variable
区域量 regional magnitude
区域量原则 regional magnitude principle
区域特征量 regional characteristic
区域特征量原则 regional characteristic principle
全域 universe
确定性地质统计学 deterministic geostatistics
确认 validation
软数据 soft data
弱平稳性 weak stationarity
三参数对数正态分布 three-parameter lognormal distribution
散点图 cross-plot; scattergram; scatterplot
熵 entropy
上四分位数 upper quartile
射影定理 projection theorem
剩余 residual
剩余克里格法 residual kriging
十分位数 decile
时间序列 time series
实现 realization
实验 experiment
实验半变差函数(图) experimental semivariogram
实验协方差 experimental covariance
事件 event
事件空间 event space
试验 trial
数 datum
数积 scalar product
数与矩阵的积 product of a scalar and a matrix
数据 data
双峰分布 bimodal distribution
双射 bijection
四分位偏差 quartile deviation
四分位数 quartile
四分位数间距 interquartile range
斯皮尔曼相关系数 Spearman correlation coefficient
似然函数 likelihood function
算术平均值 arithmetic average; arithmetic mean
算子 operator
随机变量 random variable; stochastic variable; aleatory variable
随机场 random field
随机过程 random process; stochastic process
随机函数 random function
随机集 random set
随机记号法 random tokens method
随机克里格法 random kriging
随机试验 random experiment
随机无关 stochastic independence
随机相关 stochastic dependence
随机掷硬币法 random coin method
损失函数 loss function
套合半变差函数(图) nested semivariogram
套合结构 intermeshed structure; nested structure

特征向量 eigenvector
特征值 eigenvalue
特异值 outlier
替森多边形 Thiessen polygon
调和的 harmonic
调和平均值 harmonic mean
条件分布 conditional distribution
条件概率 conditional probability
条件概率密度函数 conditional probability density function
条件模拟 conditional simulation
条件偏倚 conditional bias
条件期望 conditional expectation
条件无偏 conditional nonbias
条件无偏性 conditional unbiasedness
条件协模拟 conditional cosimulation
同方差的 homoscedastic
同位的 isotopic
统计量 statistic
统计无关 statistical independence
统计相关 statistical dependence
凸函数 convex function
椭园各向异性 elliptical anisotropy
外漂移法 external drift method
外推 extrapolation
外延方差 extension variance
完全空间 complete space
网格 grid
微遍历性 microergodicity
维格纳-塞茨格子 Wigner–Seitz cell
维纳－勒维过程 Wiener–Levy process
维纳过程 Wiener process
维纳滤波 Wiener filter
稳定的 resistant
稳健的 robust
沃罗诺瓦多边形 Voronoi polygon
沃罗诺瓦剖分 Voronoi division
沃罗诺瓦嵌形 Voronoi tesselation
五次球状半变差函数(图) penta-spherical semivariogram
无偏的 unbiased
无限可分性 infinite divisibility
误差 error
西切尔 t 分布 Sichel t distribution
希尔伯特空间 Hilbert space
析取克里格法 disjunctive kriging
下确界 infimum
下四分位数 lower quartile
先验方差 a priori variance
先验分布 prior distribution
线性半变差函数(图) linear semivariogram
线性地质统计学 linear geostatistics
线性空间 linear space
线性流形 linear manifold
线性滤波 linear filtering
线性无关 linear independence
线性相关 linear dependence
相对半变差函数(图) relative semivariogram
相对块金效应 relative nugget effect
相关 correlation
相关图 correlogram
相关系数 correlation coefficient
相似性维数 similarity dimension
镶嵌模型 mosaic model
向量 vector
向量空间 vector space
协方差 covariance
协方差矩阵 covariance matrix
协克里格法 cokriging
协模拟 cosimulation
协区域化 coregionalization
协区域化的线性模型 linear model of coregionalization
协条件模拟 coconditional simulation
信息效应 information effect
许瓦尔兹不等式 Schwarz inequality
序列变异函数 serial variation function
序列指标模拟 sequential indicator simulation
旋转带法 turning bands method
选择性的 selective
选择性曲线 selectivity curve
严格平稳性 strict stationarity
严格条件正定函数 strictly conditionally positive definite function
严格正定函数 strictly positive definite function
样本 sample
样本函数 sample function
样本空间 sample space
样本描述空间 sample description space
样本容量 sample size
样品 sample
样条函数 spline function
一般相对半变差函数(图) general relative semivariogram
一阶变差函数 madogram
一致条件化 uniform conditioning
一致性 consistency
一元的 univariate
移动线性平均 moving linear average
异方差的 heteroscedastic
异位的 heterotopic
因子分析 factor analysis
因子克里格法 factorial kriging
影响带 zone of influence
影响多边形 polygon of influence
硬数据 hard data
有理表达式 rational expression
有限阶平稳的 stationary of finite order
域 domain
预测 prediction
预后验分布 pre-posterior distribution
原地储量 reserves in place
原点处的表观不连续性 apparent discontinuity at the origin
约定参数 conventional parameter

约化剩余 reduced residual
约束区间 constraint interval
约束效应 constraint effect
跃迁的 transitive
跃迁法 transitive method
跃迁克里格法 transitive kriging
跃迁协变差函数 transitive covariogram
允许系数的平移 translation of permissible coefficients
允许性准则 permissibility criteria
余维数 codimension
整体 universe
整体的 global
整体估计 global estimation
正问题 normal problem; direct problem; forward problem
正定 positive definite
正规方程组 normal equations
正交多项式 orthogonal polynomial
正交多项式的完全集 complete set of orthogonal polynomials
正交函数 orthogonal function
正交函数的完全集 complete set of orthogonal functions
正交随机过程 orthogonal random process
正交随机函数 orthogonal random function
正交向量 orthogonal vector
正态变形 Gaussian anamorphosis
正态变换 normal score transform
正态得分变换 normal score transform
正态分布 normal distribution
正则化 regularization
正则化半变差函数(图) regularized semivariogram
正则系统 regular system
支撑 support
支撑效应 support effect
直方图 histogram
指标 indicator
指标互协方差 indicator cross-covariance
指标克里格法 indicator kriging
指标数据列 indicator data column
指标协方差 indicator covariance
指标协克里格法 indicator cokriging
指标主分量克里格法 indicator principal components kriging
指数半变差函数(图) exponential semivariogram
指数半变差函数模型 power semivariogram model
指数分布 exponential distribution
滞后距 lag
置信区间 confidence interval
置信限 confidence limit
中位数 median
中位数指标克里格法 median indicator kriging
中心极限定理 central limit theorem
中心协方差 centered covariance
众数 mode
主分量 principal component
主元 pivot
转换函数 transfer function
转移概率 transition probability
准平稳的 quasi-stationary
准确度 accuracy
自克里格法 autokriging
自克里格性 autokrigeability
自相关 autocorrelation
自相关函数 autocorrelation function
自相关图 autocorrelogram
自相关系数 autocorrelation coefficient
自相似 selfsimilar
自协方差 autocovariance
秩序变差函数(图) rank order variogram
秩序变换 rank order transform
总体 population; ensemble
最小二乘法 least squares method
最优剩余 optimal residual
摺刀法 jackknife

FRENCH to ENGLISH

σ–algèbre. σ–algebra.
accumulation. accumulation.
affinité. affinity transformation.
ajustement de courbe. curve fitting.
ajustement du variogramme. semivariogram fitting.
analyse de Fourier. Fourier analysis.
analyse factorielle. factor analysis.
analyse krigeante. factorial kriging.
analyse structurale. structural analysis.
anamorphose. anamorphosis; normal score transform.
anamorphose gaussienne. Gaussian anamorphosis.
anisotropie. anisotropy.
anisotropie elliptique. elliptical anisotropy.
anisotropie géométrique. geometric anisotropy.
anisotropie zonale. stratified anisotropy; zonal anisotropy.
aplatissement. kurtosis.
autocorrelation. autocorrelation.
autocorrélogramme. autocorrelogram.
autocovariance. autocovariance.
auto-homothétique. selfsimilar.
autokrigeabilité. autokrigeability.
autokrigeage. autokriging.
biais. bias.
biais conditionnel. conditional bias.
bijection. bijection.
bivariable. bivariate.
bruit blanc. white noise.
capacité de Choquet. Choquet's capacity.
caractéristique intrinsèque. intrinsic attribute.
cellule de Meijering. Meijering's cell.
cellule de Wigner-Seitz. Wigner–Seitz cell.
centile. centile.
champ. domain; field.
champ aléatoire. random field.
champ aléatoire d'ordre 2. second order random field.
champ Borelien. Borel field.
champ géométrique. geometric field.
codimension. codimension.
coefficient d'aplatissement. coefficient of kurtosis.
coefficient d'asymétrie. coefficient of skewness.
coefficient d'autocorrélation. autocorrelation coefficient.
coefficient de corrélation. correlation coefficient.
coefficient de correlation croisée. cross-correlation coefficient.
coefficient de correlation de Pearson. Pearson correlation coefficient.
coefficient de correlation de Spearman. Spearman correlation coefficient.
coefficient de variation. coefficient of variation.
coefficient d'excès. coefficient of excess.
coefficients autorisés d'ordre υ. authorized coefficients of order υ; permissible coefficients of order υ.
cofacteur d'un élément d'un déterminant. cofactor of an element in a determinant.
cohérence. consistency.
cokrigeage. cokriging.
cokrigeage d'indicatrices. indicator cokriging.
colonne des données d'indicatrice. indicator data column.
colonne d'un déterminant. column of a determinant.
colonne d'une matrice. column of a matrix.
combinaison linéaire de variables. linear combination of variables.
combinaisons autorisées d'ordre υ . authorized combinations of order υ; permissible combinations of order υ.
compact. compact.
comportement parabolique. parabolic behavior.
composante principale. principal component.
concept opératoire. operational concept.
condition d'universalité. universal condition.
conditionnement uniforme. uniform conditioning.
configuration. configuration.
convolution. convolution.
corégionalisation. coregionalization.
correction affine. affine correction.
correction lognormale. lognormal correction.
correction lognormale indirecte. indirect lognormal correction.
corrélation. correlation.
corrélation croisée. cross-correlation.
corrélation intrinsèque. intrinsic correlation.
corrélation spatiale. spatial dependence.
corrélogramme. correlogram.
corrélogramme croisé. crosscorrelogram.
cosimulation. cosimulation.
cosimulation conditionnelle. coconditional simulation; conditional cosimulation.
cote centré reduit. z-score.
courbe de fréquence. frequency curve.
courbe de niveau. isoline; isopleth.
courbe de sélectivité. selectivity curve.
covariance. covariance.
covariance centrée. centered covariance.
covariance croisée. cross-covariance.
covariance d'indicatrice. indicator covariance.
covariance d'indicatrice croisée. indicator cross-covariance.

covariance expérimentale. experimental covariance.
covariance expérimentale nonergodique. nonergodic experimental covariance.
covariance generalisée d'ordre υ. generalized covariance of order υ.
covariance non centrée. noncentered covariance.
covariogramme géométrique. geometric covariogram.
covariogramme transitif. transitive covariogram.
critère de cohérence. permissibility criteria.
critère d'objectivité. objectivity criterion.
decile. decile.
décomposition L.U. lower-upper (LU) decomposition.
déconvolution. deconvolution.
défini positif. positive definite.
défini positif conditionnel d'ordre υ. conditionally positive definite of order υ.
dégroupement. declustering.
demi-variogramme. demivariogram; semivariogram.
demi-variogramme croisé. cross-semivariogram.
demi-variogramme relatif general. general relative semivariogram.
demi-variogramme relatif local. local relative semivariogram.
demi-variogramme relatif par paire. pairwise relative semivariogram.
densité bivariable isofactorielle. diagonal bivariate density function; isofactorial bivariate density function.
densité marginale. marginal probability density function.
densité multivariable. joint probability density function.
dependance linéaire. linear dependence.
dépendance statistique. statistical dependence.
dépendance stochastique. stochastic dependence.
derive. drift.
déstructuration des hautes teneurs. high-grade destructuration.
déterminant. determinant.
déterminant d'une matrice. determinant of a matrix.
déviation quartile. quartile deviation.
diagonale principale d'une matrice. main diagonal of a matrix; principal diagonal of a matrix.
diagramme de frequence. frequency diagram.
dimension de Hausdorff-Besicovitch. Hausdorff–Besicovitch dimension
dimension de similitude. similarity dimension.
dimension fractale. fractal dimension.
discontinuité apparente à l'origine. apparent discontinuity at the origin.
dispersion absolue. absolute dispersion.
dispersion intrinsèque. intrinsic dispersion.
distribution. distribution.
distribution bimodale. bimodal distribution.
distribution de probabilité. probability distribution
distribution marginale. marginal distribution.
distribution monomodale. unimodal distribution.
distribution multimodale. multimodal distribution.
disymétrie. skewness.
domaine. domain.
donnée. datum.
donneé indicatrice floue. fuzzy indicator data.
donnée dure. hard data.
donnée molle. soft data.
données. data.
données aberrantes. outlier.
ecart-type. standard deviation.
échantillon. sample.
échantillonnage aléatoire stratifié. stratified random sampling.
échelle de travail. working scale.
effet d'ecran. screen effect.
effet de contrainte. constraint effect.
effet de pépite. nugget effect.
effet de pépite pur. pure nugget effect.
effet de pépite relatif. relative nugget effect.
effet de support. support effect.
effet de trou. hole effect.
effet d'information. information effect.
effet proportionnel. proportional effect.
élément d'un déterminant. element of a determinant.
élément d'une matrice. element of a matrix.
ellipse de Krige. Krige's ellipse.
ensemble. ensemble.
ensemble aléatoire. random set.
entropie. entropy.
équation caractéristique d'une matrice. characteristic equation of a matrix.
équations normales. normal equations.
ergodicité. ergodicity.
erreur. error.
erreur quadratique moyenne. mean square error.
erreur-type. standard error.
espace complet. complete space.
espace de Hilbert. Hilbert space.
espace de probabilité. probability space.
espace de produit scalaire. inner product space.
espace des échantillons. event space; sample space.
espace des états d'un système. sample description space.
espace euclidien. euclidean space.
espace métrique. metric space.
espace vectoriel. linear space; vector space.
espérance conditionnelle. conditional expectation.
espérance mathématique. expected value; mean.
estimateur. estimator.
estimateur de moindre erreur carrée. minimum mean square error estimator.
estimateur de type E. E-type estimate.
estimateur du maximum de vraisemblance. maximum likelihood estimator; maximum likelihood statistic.
estimateur minimisant la norme de l'error. mini-

mum error norm estimator.
estimation. estimation.
estimation globale. global estimation.
estimation locale. local estimation.
étendue. range.
évènement. event.
exactitude. accuracy.
expérience. experiment.
expérience aléatoire. random experiment.
expression rationelle. rational expression.
extrapolation. extrapolation.
facteur de correction de variance. variance adjustment factor.
fenêtre. window.
fermé. closed.
filtrage. filtering.
filtrage linéaire. linear filtering.
filtre de Wiener. Wiener filter.
fonction. function.
fonction aléatoire. random function.
fonction aléatoire booléenne. Boolean random function.
fonction aléatoire en bruit blanc. white random function.
fonction aléatoire factorisable. factorable random function.
fonction aléatoire intrinsèque d'ordre υ . intrinsic random function of order υ.
fonction aléatoire orthogonale. orthogonal random function.
fonction aléatoire semi-markovienne. semimarkovian random function.
fonction aléatoire simultanément orthogonal. jointly orthogonal random function.
fonction aléatoires simultanément factorisables. jointly factorable random function.
fonction Beta. beta function.
fonction convexe. convex function.
fonction d'autocorrélation. autocorrelation function.
fonction de base. basis function.
fonction d'échantillonnage. sample function.
fonction de densité. probability density function.
fonction de densité conditionnelle. conditional probability density function.
fonction d'efficacité. objective function.
fonction de fréquence. frequency function.
fonction d'ensemble additive. additive set function.
fonction de perte. loss function.
fonction de probabilité. probability function.
fonction de récupération. recovery function.
fonction de répartition. cumulative distribution function; distribution function.
fonction de structure. structure function.
fonction de transfert. transfer function.
fonction de vraisemblance. likelihood function.
fonction d'Heaviside. Heaviside's function.
fonction échelon. step function.
fonction Gamma. gamma function.
fonction intégrable. integrable function.
fonction intégrable par rapport à une deuxième fonction. integrable function with respect to a second function.
fonction intrinsèque. intrinsic function.
fonction orthonormales. orthonormal function.
fonctions orthogonales. orthogonal function.
fonction spline. spline function.
fonction strictement définie positive. strictly positive definite function.
fonction strictement définie positive conditionnelle. strictly conditionally positive definite function.
fonction structurale. structural function.
formalisme de Lagrange. Lagrange's method of multipliers.
formalisme du maximum entropie. maximum entropy formalism.
formule de de Wijs. de Wijs formula.
formule de Krige. Krige's formula.
formule de Parseval. Parseval's equation.
formule de Rodrigues. Rodrigues formula.
fractal. fractal.
fractile quantile. fractile.
frequence cumulée. cumulative frequency.
géostatistique. geostatistics.
géostatistique déterministe. deterministic geostatistics.
géostatistique linéaire. linear geostatistics.
géostatistique non linéaire. nonlinear geostatistics.
géostatistique non paramètrique. nonparametric geostatistics.
global. global.
grandeur régionale. regional characteristic; regional magnitude
graphe de dispersion. scatterplot.
graphe en coordonées polaires. rose diagram.
grille. grid.
harmonique. harmonic.
hétéroscédastique. heteroscedastic.
hétérotopique. heterotopic.
histogramme. histogram.
homogénéité. homogeneity.
homoscedastique. homoscedastic.
hypothèse de permanence. conservation of law hypothesis.
hypothèse intrisèque. intrinsic hypothesis.
impulsion. impulse.
incertitude. uncertainty.
increment généralisé d'ordre υ. generalized increments of order υ.
indépendance linéaire. linear independence.
indépendance statistique. statistical independence.
indépendance stochastique. stochastic independence.
indicatrice. indicator.

inegalité de Buniakovski. Buniakovski inequality.
inegalité de Cauchy-Schwarz. Cauchy–Schwarz inequality.
inégalité de Bessel. Bessel inequality.
inégalité de Schwarz. Schwarz inequality.
infimum. infimum.
infinie divisibilité. infinite divisibility.
infinie divisibilité pour la réunion. infinite divisibility of a union.
intérieur. interior.
interpolation. interpolation.
interpolateur exacte. exact interpolator.
interquartile. interquartile range.
intervalle de confiance. confidence interval.
intervalle de contrainte. constraint interval.
intervalle de probabilité. probability interval.
intrinsèque. intrinsic.
isotopique. isotopic.
isotropie. isotropy.
Jackknife. jackknife.
krigeage. kriging.
krigeage aléatoire. random kriging.
krigeage de blocs. block kriging.
krigeage de indicatrice de médiane. median indicator kriging.
krigeage de probabilité. probability kriging.
krigeage des composantes principales d'indicatrices. indicator principal component kriging.
krigeage des résidus. residual kriging.
krigeage d'indicatrice multiple. multiple indicator kriging.
krigeage d'indicatrice unique. single indicator kriging.
krigeage d'indicatrice. indicator kriging.
krigeage disjonctif. disjunctive kriging.
krigeage dual. dual kriging.
krigeage intrinsèque. intrinsic kriging.
krigeage logarithmique. logarithmic kriging.
krigeage lognormal. lognormal kriging.
krigeage multigaussien. multigaussian kriging.
krigeage ordinaire. ordinary kriging.
krigeage ponctuel. point kriging; punctual kriging.
krigeage simple. simple kriging.
krigeage transitif. transitive kriging.
krigeage universel. universal kriging.
leptokurtique. leptokurtic.
ligne de contour. contour.
ligne de régression. regression line.
ligne d'un déterminant. row of a determinant.
ligne d'une matrice. row of a matrix.
limite de confiance. confidence limit.
lissage. smoothing.
local. local.
loi a priori. prior distribution.
loi Beta. beta distribution.
loi bigaussienne. bigaussian model.
loi binomiale. binomial distribution.
loi binomiale négative. negative binomial distribution.
loi bivariable. bivariate distribution.
loi conditionelle. posterior distribution; conditional distribution.
loi de Gauss. Gaussian distribution.
loi de Poisson. Poisson distribution.
loi des grands nombres. law of large numbers.
loi de Sichel. Sichel t distribution.
loi exponentielle. exponential distribution.
loi Gamma. gamma distribution.
loi lognormale. lognormal distribution.
loi lognormale à deux paramètres. two-parameter lognormal distribution.
loi lognormale à trois paramètres. three-parameter lognormal distribution.
loi multigaussienne. multinormal distribution.
loi multivariable. joint probability distribution; multivariate distribution.
loi normale. bivariate normal distribution; normal distribution; standard normal distribution.
loi pre-postérieure. pre-posterior distribution.
loi uniforme. uniform distribution.
madogramme. madogram.
matrice. matrix.
matrice carrée. square matrix.
matrice des covariances. covariance matrix.
matrice diagonale. diagonal matrix.
matrice diagonalisée. diagonalized matrix.
matrice identité. identity matrix.
matrice Hermitienne. Hermitian matrix.
matrice inverse. inverse of a matrix.
matrice singulière. singular matrix.
matrice symétrique. symmetric matrix.
matrice transposée. transpose of a matrix.
médiane. median.
mésocurtique. mesokurtic.
mesure. measure.
mesure de Dirac. Dirac delta function.
mesure de Hausdorff. Hausdorff measure.
mesure de probabilité. probability measure.
méthode de Choleski. Choleski's method.
méthode de la dérive externe. external drift method.
méthode de Monte-Carlo. Monte Carlo method.
méthode des bandes tournantes. turning bands method.
méthode des jetons aléatoires. random tokens method; random coin method.
méthode des moindres carrés. least squares method.
méthode transitive. transitive method.
microergodicité. microergodicity.
mode. mode.
modèle. model.
modèle d'effet de pépite. pure nugget effect semivariogram model.
modèle de changement de support. change of sup-

port model.
modèle de variogramme à effet de trou. hole effect semivariogram model.
modèle gaussien discrétisé. discrete Gaussian model.
modèle hermitien. Hermite model.
modèle linéaire de corégionalisation. linear model of coregionalization.
modèle mosaïque. mosaic model.
modèle topo-probabiliste. topoprobabilistic model.
moment. moment.
moment d'une loi. moment of a distribution.
monoscopique. monoscopic.
montée. grading.
mosaique de Dirichlet. Dirichlet tesselation.
mosaique de Voronoi. Voronoi tesselation.
mouvement Brownien. Brownian motion process.
moyenne. average value; average.
moyenne arithmétique. arithmetic average; arithmetic mean.
moyenne géométrique. geometric average; geometric mean.
moyenne harmonique. harmonic mean.
moyenne mobile. moving linear average.
multiplicateur de Lagrange. Lagrange multiplier.
multivariable. multivariate.
nombre imaginaire conjugué. conjugate complex number.
nonbiais conditionnel. conditional nonbias; conditional unbiasedness.
noncorrélation spatiale. spatial independence.
norme. norm.
note réduit. standard score
nuage de corrélation. cross-plot; scattergram.
nuage de corrélation différée. h-scattergram.
opérateur. operator.
ordre d'un déterminant. order of a determinant.
ordre d'une matrice carrée. order of a square matrix.
ouvert. open.
palier. sill.
panscopique. panscopic.
paramètre conventionnel. conventional parameter.
partition de Delaunay. Delaunay subdivision.
partition de Voronoi. Voronoi division.
partition d'un ensemble. partition of a set.
percentile. percentile.
pivot. pivot.
platycurtique. platykurtic.
point d'accumulation. accumulation point.
polygone d'influence. polygon of influence.
polygone de Thiessen. Thiessen polygon.
polygone de Voronoi. proximal polygon; Voronoi polygon.
polynôme a plusieurs variables. polynomial in several variables.
polynôme de Laguerre. Laguerre polynomials.
polynôme d'ordre *n*. polynomial of degree *n*.
polynômes d'Hermite. Hermite polynomials.
polynômes orthogonaux. orthogonal polynomial.
polyscopique. polyscopic.
population. population; universe.
portée. range.
portée intégral. integral range.
précision. precision.
prédiction. prediction.
principe des grandeurs régionales. regional characteristic principle; regional magnitude principle.
probabilité. probability.
probabilité conditionnelle. conditional probability.
probabilité de transition. transition probability.
problème direct. direct problem; forward problem; normal problem.
problème inverse. inverse problem.
processus aléatoire. random process.
processus aléatoire orthogonal. orthogonal random process.
processus aléatoire simultanément orthogonal. jointly orthogonal random process.
processus aléatoire factorisable. factorable random process.
processus de Markov. Markov process.
processus de Poisson. Poisson process.
processus de Wiener. Wiener process.
processus de Wiener-Lévy. Wiener–Levy process.
processus stochastique. stochastic process.
produit d'une matrice par un scalaire. product of a scalar and a matrix.
produit de matrices. product of matrices.
produit scalaire. inner product; scalar product.
propriété ergodique. ergodic property.
q-q graphe. q-q plot.
quantile. quantile.
quartile. quartile.
quartile inférieur. lower quartile.
quartile supérieur. upper quartile.
quasi-stationnaire. quasi-stationary.
réalisation. outcome; realization.
récursivité. recursion.
régression. regression.
régularisation. regularization.
relation de Cartier. Cartier's relation.
représentation d'une fonction aléatoire intrinsèque. representation of an intrinsic random function.
représentation glissante. sliding representation.
réserves in situ. reserves in place.
réserves récupérables. recoverable reserves.
résidu. residual.
résidu optimal. optimal residual.
résidu standard. reduced residual.
resistant. resistant.
retard. lag.
robustesse. robust.
rodogramme. rodogram.

sans biais. unbiased.
schéma booléen. Boolean random function.
schéma de De Wijs. de Wijsian semivariogram.
schéma de dilution. dilution random function.
sélectif. selective.
sens objectif. objective meaning.
serial variation function. serial variation function.
série temporelle. time series.
sigma-algèbre. sigma-field.
signification objective. objective meaning.
simulation conditionnelle. conditional simulation.
simulation sequentielle d'indicatrices. sequential indicator simulation.
simulation. simulation.
sous-estimation. underestimation.
stabilité pour la réunion. union stability.
stationnaire. stationary.
stationnaire au sens large. stationary in the wide sense.
stationnarité. stationarity.
stationnarité d'ordre fini. stationary of finite order.
stationnarité faible. weak stationarity.
stationnarité stricte. strict stationarity; strong stationarity.
statistique. statistic.
structuration des hautes teneurs. high-grade structuration.
structure. structure.
structures gigognes. intermeshed structure; nested structure.
suite de Cauchy. Cauchy sequence.
support. support.
support géométrique. geometric support.
surestimation. overestimation.
symbole de Kronecker. Kronecker delta.
système nonsingulier. nonsingular system.
système complet de fonctions orthogonales. complete set of orthogonal functions.
système complet de polynômes orthogonaux. complete set of orthogonal polynomials.
système régulier. regular system.
taille de l'énchantillon. sample size.
tendance régionale. trend.
théorème d'additivité. additivity theorem.
théorème de Bayes. Bayes' theorem.
théorème de la limite centrale. central limit theorem.
théorème de projection. projection theorem.
théorème de Pythagore. Pythagorean theorem.
théorème ergodique. ergodic theorem.
théorie. theory.
tirage. trial.
tirages de Bernoulli. Bernoulli experiment.
transformation. transformation.
transformation de Fourier. Fourier transform.
transformation de Fourier inverse. Fourier synthesis.
transformation de Fourier rapide. fast Fourier transform.
transformation de Laplace. Laplace transform.
transformation spatiale de classe 1. space transformation of the first kind.
transformation spatiale de classe 2. space transformation of the second kind.
transformée. transform.
transformée de rang d'ordre. rank order transform.
transformée uniforme. uniform transformation.
transitif. transitive.
translation de coefficients autorisés. translation of permissible coefficients.
triangulation de Delaunay. Delaunay triangulation.
univariable. univariate.
universel. universal.
valeur estimée. estimate.
valeur propre. eigenvalue.
valeur propre d'une matrice. eigenvalue of a matrix.
validation. validation.
validation croisée. cross-validation.
validation croisée multiple. joint cross-validation.
variable aléatoire. aleatory variable; random variable; variate.
variable aléatoire continue. continuous random variable.
variable aléatoire discrète. discrete random variable.
variable aléatoire d'ordre 2. second order random variable.
variable régionalisée. regionalized variable.
variable stochastique. stochastic variable.
variance. variance.
variance a priori. a priori variance.
variance de krigeage. kriging variance.
variance de dispersion. dispersion variance.
variance d'estimation. estimation variance.
variance d'extension. extension variance.
variété. manifold.
variété linéaire. linear manifold.
variogramme. variogram.
variogramme régional. regional semivariogram.
variogramme croisé. cross-variogram.
variogramme cubique. cubic semivariogram.
variogramme de rang d'ordre. rank order variogram.
variogramme d'ordre α. variogram of order α.
variogramme expérimental. experimental semivariogram.
variogramme exponentiel. exponential semivariogram.
variogramme gaussien. Gaussian semivariogram.
variogramme généralisé. generalized semivariogram.
variogramme gigogne. nested semivariogram.
variogramme linéaire. linear semivariogram.
variogramme logarithmique. logarithmic semivariogram.
variogramme modélisé. semivariogram model.
variogramme monté. graded semivariogram.

variogramme moyen. average semivariogram; F-function; mean semivariogram.
variogramme parabolique. parabolic semivariogram.
variogramme pentasphérique. penta-spherical semivariogram.
variogramme puissance. power semivariogram model.
variogramme relatif. relative semivariogram.
variogramme régional probabilisé. probabilistic regional semivariogram.
variogramme régularisé. regularized semivariogram.
variogramme sous-jacent. underlying semivariogram.
variogramme sphérique. spherical semivariogram.
variogramme théorique. theoretical semivariogram.
variographie. variography.
vecteur. vector.
vecteur propre. eigenvector.
vecteurs orthogonaux. orthogonal vector.
vecteurs orthonormaux. orthonormal vector.
voisinage. neighborhood.
voisinage de krigeage. kriging neighborhood.
voisinage de travail. working neighborhood.
voisinage glissant. sliding neighborhood.
zone d'influence. zone of influence.

GERMAN to ENGLISH

σ–Algebra. σ–algebra.
abgeschlossen. closed.
Abschirmeffekt. screen effect.
absolute Dispersion. absolute dispersion.
Abstand. lag.
Additionssatz, -theorem. additivity theorem.
additive Mengenfunktion. additive set function.
affine Korrektur. affine correction.
affine Transformation. affinity transformation.
Akkumulation. accumulation.
allgemeines relatives Semivariogramm. general relative semivariogram.
Anamorphismus. anamorphosis; normal score transform.
Anamorphose. anamorphosis.
Anisotropie. anisotropy.
Anpassung eines Variogramms. semivariogram fitting.
a-posteriori Verteilung. posterior distribution.
a-priori-Varianz. a priori variance.
a-priori Verteilung. prior distribution.
Arbeitsbereich. working neighborhood.
Arbeitsmaßstab. working scale.
arithmetisches Mittel. arithmetic average; arithmetic mean.
Ausbringungsfunktion. recovery function.
Ausdehnungsvarainz. extension variance.
Ausreißer. outlier.
Autokorrelation. autocorrelation.
Autokorrelationsfunktion. autocorrelation function.
Autokorrelationskoeffizient. autocorrelation coefficient.
Autokorrelogram. autocorrelogram.
Autokovarianz. autocovariance.
Autokrigbarkeit. autokrigeability.
Autokrigen. autokriging.
Basisfunktion. basis function.
bedingte Erwartung. conditional expectation.
bedingte Kosimulation. conditional cosimulation; coconditional simulation.
bedingte Simulation. conditional simulation.
bedingte Unverzerrtheit. conditional nonbias; conditional unbiasedness.
bedingte Verteilung. conditional distribution.
bedingte Verzerrung. conditional bias.
bedingte Wahrscheinlichkeit. conditional probability.
bedingte Wahrscheinlichkeitsdichte. conditional probability density function.
bedingt positiv definit der Ordnung v. conditionally positive definite of order v.
begrenzendes Intervall. constraint interval.
Bernoulli-Versuch. Bernoulli experiment.
Besselsche Ungleichung. Bessel inequality.
Betafunktion. beta function.
Betaverteilung. beta distribution.
Bezugsgröße. datum (a)
bijektive Abbildung. bijection.
bimodale Verteilung. bimodal distribution.
Binomialverteilung. binomial distribution.
Block-Krigen. block kriging.
Boolesche Zufallsfunktion. Boolean random function.
Borel-Feld. Borel field.
Brownsche Bewegung. Brownian motion process.
Buniakovskische Ungleichung. Buniakovski inequality.
Cartiersche Formel. Cartier's relation.
Cauchy-Folge. Cauchy sequence.
Cauchy–Schwarzsche Ungleichung. Cauchy–Schwarz inequality.
charakteristische Gleichung einer Matrix. characteristic equation of a matrix.
Choleskische Methode. Choleski's method.
Choquetsche Eigenschaft. Choquet's capacity.
Darstellung einer intrinsischen Zufallsfunktion. representation of an intrinsic random function.
das Innere. interior.
Daten. data.
Datum. datum (b).
Delaunay-Triangulation. Delaunay triangulation.
Delaunay-Zerlegung. Delaunay subdivision.
Destrukturierung großer Gehalte. high-grade destructuration.
Determinante. determinant
Determinante einer Matrix. determinant of a matrix.
deterministische Geostatistik. deterministic geostatistics.
de Wijssche Formel. de Wijs formula.
de Wijssches Semivariogramm. de Wijsian semivariogram.
Dezil. decile.
diagonalisierte Matrix. diagonalized matrix.
Diagonalmatrix. diagonal matrix.
Diracsche Delta-Funktion. Dirac delta function.
direkte Aufgabe. direct problem; forward problem; normal problem.
Dirichlet-Parkettierung. Dirichlet tesselation.
disjunktives Krigen. disjunctive kriging.
diskretes Gaußsches Modell. discrete Gaussian model.
diskrete Zufallsvariable. discrete random variable.

Dispersionsvarianz. dispersion variance.
Domäne. domain.
dreiparametrische lognormale Verteilung. three-parameter lognormal distribution
Drift. drift.
duales Krigen. dual kriging.
Effekt der Rahmenbedingungen. constraint effect.
Eigenvektor. eigenvector.
Eigenwert. eigenvalue.
Eigenwert einer Matrix. eigenvalue of a matrix.
eindimensional. univariate.
eineindeutige Abbildung. bijection.
einfaches Indikatorkrigen. single indicator kriging.
einfaches Krigen. simple kriging.
Einflußbereich. zone of influence.
Einflußpolygon. polygon of influence.
Einheitsmatrix. identity matrix.
Element einer Matrix. element of a matrix.
Element in einer Determinante. element of a determinant.
elliptische Anisotropie. elliptical anisotropy.
Ensemble. ensemble.
Entclusterung. declustering.
Entfaltung. deconvolution.
Entropie. entropy.
Ereignis. event.
Ereignisraum. event space.
Ergodeneigenschaft. ergodic property.
Ergodensatz. ergodic theorem.
Ergodizität. ergodicity.
Erwartungswert. expected value.
Euklidischer Raum. euclidean space.
Exaktheit. precision.
Experiment. experiment.
experimentelle Kovarianz. experimental covariance.
experimentelles Semivariogramm. experimental semivariogram.
Exponentialverteilung. exponential distribution.
exponentielles Semivariogramm. exponential semivariogram.
Extrapolation. extrapolation.
Exzeß. coefficient of excess.
Faktorenanalyse. factor analysis.
faktorielles Krigen. factorial kriging.
faktorisierbarer Zufallsprozeß. factorable random process.
faktorisierbare Zufallsfunktion. factorable random function.
Faltung. convolution.
Fehler. error.
Feld. domain; field.
Fenster. window.
F-Funktion. F-function.
filtern. filtering.
flachgipflig. platykurtic.
Fourieranalyse. Fourier analysis.
Fouriersynthese. Fourier synthesis.
Fouriertransformierte. Fourier transform.
Fraktal. fractal.
fraktale Dimension. fractal dimension; similarity dimension.
Fraktil. fractile.
Funktion. function.
Fuzzy Indikatordaten. fuzzy indicator data.
Gammafunktion. gamma function.
Gammaverteilung. gamma distribution.
Gaußscher Anamorphismus. Gaussian anamorphosis.
Gaußsches Semivariogramm. Gaussian semivariogram.
Gaußsche Verteilung. Gaussian distribution.
gemeinsame Kreuzprüfung. joint cross-validation.
gemeinsame Wahrscheinlichkeitsdichte. joint probability density function.
gemeinsame Wahrscheinlichkeitsfunktion. joint probability distribution.
gemeinsam faktorisierbare Zufallsfunktion. jointly factorable random function.
gemeinsam orthogonaler Zufallsprozeß. jointly orthogonal random process.
gemeinsam orthogonale Zufallsfunktion. jointly orthogonal random function.
Genauigkeit. accuracy.
geometrische Anisotropie. geometric anisotropy.
geometrisches Feld. geometric field.
geometrisches Kovariogram. geometric covariogram.
geometrisches Mittel. geometric average; geometric mean.
geometrische Stützung. geometric support.
Geostatistik. geostatistics.
Gesamtheit. ensemble; population; universe.
geschachteltes Semivariogramm. nested semivariogram.
geschachtelte Struktur. intermeshed structure; nested structure.
geschichtet zufällige Probennahme. stratified random sampling.
Gesetz der großen Zahlen. law of large numbers.
gewinnbare Reserven. recoverable reserves.
Gitter. grid.
Glättung. smoothing.
gleichmäßiges Bedingen. uniform conditioning.
gleichmäßige Verteilung. uniform distribution.
Gleichverteilung. uniform distribution.
gleitende Darstellung. sliding representation.
gleitende Nachbarschaft. sliding neighborhood.
gleitendes Mittel. moving linear average.
global. global.
globale Schätzung. global estimation.
globale Vorräte. reserves in place.
gradieren. grading.
gradiertes Semivariogramm. graded semivariogram.
Gültigkeitstest. validation.

harmonisch. harmonic.
harmonisches Mittel. harmonic mean.
harte Daten. hard data.
Häufigkeitsdiagram. frequency diagram.
Häufigkeitsfunktion. frequency function.
Häufigkeitskurve. frequency curve.
häufigster Wert. mode.
Häufungspunkt. accumulation point.
Hauptdiagonale einer Matrix. main diagonal of a matrix; principal diagonal of a matrix.
Hauptkomponente. principal component.
Hausdorff–Besicovitch-Dimension. Hausdorff–Besicovitch dimension.
Hausdorff-Maß. Hausdorff measure.
Heavisidesche Funktion. Heaviside's function.
Hermitesche Matrix. Hermitian matrix.
Hermitesche Polynome. Hermite polynomials.
Hermitesches Modell. Hermite model.
heteroskedastisch. heteroscedastic.
heterotop. heterotopic.
Hilbert-Raum. Hilbert space.
Histogramm. histogram.
Homogenität. homogeneity.
homoskedastisch. homoscedastic.
h-Streudiagramm. h-scattergram.
Hypothese der Erhaltung. conservation of law hypothesis.
Impuls. impulse.
Indikator. indicator.
Indikatordatenspalte. indicator data column.
Indikatorkokrigen. indicator cokriging.
Indikatorkovarianz. indicator covariance.
Indikatorkrigen. indicator kriging.
indirekte lognormale Korrektur. indirect lognormal correction.
Infimum. infimum.
Informationseffekt. information effect.
inneres Produkt. inner product.
in-situ Vorräte. reserves in place.
Integrationsbereich. integral range.
integrierbare Funktion. integrable function.
integrierbare Funktion in Bezug auf eine zweite Funktion. integrable function with respect to a second function.
Interpolation. interpolation.
Interpolator im engeren Sinne. exact interpolator.
Intervall zwischen dem unteren und oberen Quartil. interquartile range.
intrinsisch. intrinsic.
intrinsische Dispersion. intrinsic dispersion.
intrinsische Funktion. intrinsic function.
intrinsische Hypothese. intrinsic hypothesis.
intrinsische Korrelation. intrinsic correlation.
intrinsisches Attribut. intrinsic attribute.
intrinsisches Krigen. intrinsic kriging.
intrinsische Zufallsfunktion der Ordnung ν. intrinsic random function of order ν.
inverse Aufgabe. inverse problem.
Inverse einer Matrix. inverse of a matrix.
isofaktorielle zweidimensionale Dichte. diagonal bivariate density function; isofactorial bivariate density function.
Isolinie. isoline.
Isoplethe. isopleth.
isotop. isotopic.
Isotropie. isotropy.
Jackknife. jackknife.
kodimension. codimension
Kofaktor eines Elements in einer Determinante. cofactor of an element in a determinant.
Kokrigen. cokriging.
kompakt. compact.
Konfiguration. configuration.
konjugiert komplexe Zahl. conjugate complex number.
Konsistenz. consistency.
Kontour. contour.
konvexe Funktion. convex function.
Koregionalisierung. coregionalization.
Korrelation. correlation.
Korrelationskoeffizient. correlation coefficient.
Korrelogramm. correlogram.
Kosimulation. cosimulation.
Kovarianz. covariance.
Kovarianzmatrix. covariance matrix.
Kreuzdiagramm. cross-plot.
Kreuzindikatorkovarianz. indicator cross-covariance.
Kreuzkorrelation. cross-correlation.
Kreuzkorrelationskoeffizient. cross-correlation coefficient.
Kreuzkorrelogramm. crosscorrelogram.
Kreuzkovarianz. cross-covariance.
Kreuzprüfung. cross-validation.
Kreuzsemivariogramm. cross-semivariogram.
Kreuzvariogramm. cross-variogram.
Krige-Beziehung. Krige's formula.
Krigen. kriging.
Krigenachbarschaft. kriging neighborhood.
Krigen der Indikatorhauptkomponenten. indicator principal component kriging.
Krigen der Residuen. residual kriging.
Krigesche Ellipse. Krige's ellipse.
Krigevarianz. kriging variance.
Kroneckersches delta. Kronecker delta.
kubisches Semivariogramm. cubic semivariogram.
kumulative Häufigkeit. cumulative frequency.
Kurvenanpassung. curve fitting.
Lagrangesche Multiplikatoren. Lagrange multiplier.
Lagrangesche Multiplikatoren Methode. Lagrange's method of multipliers.
Laguerresche Polynome. Laguerre polynomials.
Laplacesche Transformierte. Laplace transform.

Likelihood-Funktion. likelihood function.
lineare Abhängigkeit. linear dependence.
lineare Geostatistik. linear geostatistics.
lineare Mannigfaltigkeit. linear manifold.
linearer Raum. linear space.
lineares Filtern. linear filtering.
lineares Modell der Koregionalisierung. linear model of coregionalization.
lineares Semivariogramm. linear semivariogram.
lineare Unabhängigkeit. linear independence.
Linearkombination von Variablen. linear combination of variables.
Linie gleicher Werte. isopleth.
Locheffeekt. hole effect.
logarithmisches Krigen. logarithmic kriging.
logarithmisches Semivariogramm. logarithmic semivariogram.
lognormale Korrektur. lognormal correction.
lognormales Krigen. lognormal kriging.
lognormale Verteilung. lognormal distribution.
lokal. local.
lokale Schätzung. local estimation.
lokales relatives Variogramm. local relative semivariogram.
Madogramm. madogram.
Mannigfaltigkeit. manifold.
Markovscher Prozeß. Markov process.
Maß. measure.
Matrix. matrix.
Matrizenprodukt. product of matrices.
Maximum-Likelihood-Schätzer. maximum likelihood estimator.
Maximum-Likelihood-Statistik. maximum likelihood statistic.
mehrdimensional. multivariate.
mehrdimensionale Normalverteilung. multinormal distribution.
mehrdimensionales Polynom. polynomial in several variables.
mehrdimensionale Verteilung. multivariate distribution.
mehrfaches Indikatorkrigen. multiple indicator kriging.
Meijerings Zelle. Meijering's cell.
Methode der drehenden Bänder. turning bands method.
Methode der externen Drift. external drift method.
Methode der kleinsten Quadrate. least squares method.
Methode der maximalen Entropie. maximum entropy formalism.
Methode der zufälligen Auszählfläche. random coin method; random tokens method.
metrischer Raum. metric space.
Mikroergodizität. microergodicity.
Mittel. average.
Mittelwert. average; average value; mean.
mittlerer quadratischer Fehler. mean square error.
mittleres Semivariogramm. average semivariogram; mean semivariogram.
Modell. model.
Modell der Stützungsänderung. change of support model.
Moment. moment.
Moment einer Verteilung. moment of a distribution.
monoskopisch. monoscopic.
Monte-Carlo-Methode. Monte Carlo method.
Mosaikmodell. mosaic model.
multigaußsches Krigen. multigaussian kriging.
multimodale Verteilung. multimodal distribution.
Nachbarschaft. neighborhood.
nächstes Polygon. proximal polygon.
negative Binomialverteilung. negative binomial distribution.
nichtergodische experimentelle Kovarianz. nonergodic experimental covariance.
nichtlineare Geostatistik. nonlinear geostatistics.
nichtparametrische Geostatistik. nonparametric geostatistics.
nichtsinguläres System. nonsingular system.
nichtzentrierte Kovarianz. noncentered covariance.
Norm. norm.
Normalengleichungen. normal equations.
normales Krigen. ordinary kriging.
normalgipflig. mesokurtic.
Normalverteilung. normal distribution.
Nuggeteffekt. nugget effect.
oberes Quartil. upper quartile.
objektive Bedeutung. objective meaning.
Objektivitätskriterium. objectivity criterion.
offen. open.
Operationskonzept. operational concept.
Operator. operator.
optimale Residuen. optimal residual.
Ordnung einer Determinante. order of a determinant.
Ordnung einer quadratischen Matrix. order of a square matrix.
orthogonale Funktion. orthogonal function.
orthogonale Polynome. orthogonal polynomial.
orthogonaler Zufallsprozeß. orthogonal random process.
orthogonale Vektoren. orthogonal vector.
orthogonale Zufallsfunktion. orthogonal random function.
orthonormale Funktionen. orthonormal function.
orthonormale Vektoren. orthonormal vector.
ortsabhängige Variable. regionalized variable.
paarweise relatives Semivariogramm. pairwise relative semivariogram.
panskopisch. panscopic.
parabolisches Semivariogramm. parabolic semivariogram.

parabolisches Verhalten. parabolic behavior.
Parsevalsche Gleichung. Parseval's equation.
Pearsonscher Korrelationskoeffizient. Pearson correlation coefficient.
Permanenzhypothese. conservation of law hypothesis.
Perzentil. percentile.
Pivotmatrixelemente. pivot.
Poissonscher Prozeß. Poisson process.
Poisson-Verteilung. Poisson distribution.
Polynom *n*-ten Grades. polynomial of degree *n*.
polyskopisch. polyscopic.
positiv definit. positive definite.
Potenzen-Semivariogrammodell. power semivariogram model.
pre-posterior Verteilung. pre-posterior distribution.
Prinzip der regionalen Größen. regional magnitude principle.
Prinzip des regionalen Kennwertes. regional characteristic principle.
Probe. sample (b).
Produkt aus einem Skalar und einer Matrix. product of a scalar and a matrix.
Projektionssatz. projection theorem.
Proportionalitätseffekt. proportional effect.
Punktkrigen. point kriging; punctual kriging.
quadratische Matrix. square matrix.
Quantil. quantile.
Quantildiagramm. q-q plot.
Quartil. quartile.
quartile Abweichung. quartile deviation.
Quasistationarität. quasi-stationary.
quintisch-sphärisches Semivariogramm. penta-spherical semivariogram.
randliche Wahrscheinlichkeitsdichte. marginal probability density function.
Randverteilung. marginal distribution.
Rangordnungstransformierte. rank order transform.
Rangordnungsvariogramm. rank order variogram.
rationaler Ausdruck. rational expression.
Raum, in dem ein inneres Produkt definiert ist. inner product space.
räumliche Abhängigkeit. spatial dependence.
räumliche Unabhängigkeit. spatial independence.
Raumtransformation 1. Art. space transformation of the first kind
Raumtransformation 2. Art. space transformation of the second kind
Realisierung. outcome; realization.
reduziertes Residuum. reduced residual.
regionale Größe. regional magnitude.
regionaler Kennwert. regional characteristic.
regionales Semivariogramm. regional semivariogram.
regionales Wahrscheinlichkeitssemivariogramm. probabilistic regional semivariogram.
Regression. regression.
Regressionsgerade. regression line.
reguläres System. regular system.
Reichweite. range.
reiner Nuggeteffekt. pure nugget effect.
Rekursion. recursion.
relativer Nuggeteffekt. relative nugget effect.
relatives Semivariogramm. relative semivariogram.
Restgröße. residual.
robust. robust.
Rodogramm. rodogram.
Rodriguessche Formel. Rodrigues formula.
Rose-Diagramm. rose diagram.
Satz des Pythagoras. Pythagorean theorem.
Satz von Bayes. Bayes' theorem.
Schätzer. estimator.
Schätzer mit minimalem mittleren quadratischen Fehler. minimum mean square error estimator.
Schätzer mit minimaler Fehlernorm. minimum error norm estimator.
Schätzung. estimation.
Schätzvarianz. estimation variance.
Schätzwert. estimate.
Schätzwert vom E-typ. E-type estimate.
scheinbare Diskontinuität im Ursprung. apparent discontinuity at the origin.
Schiefe. coefficient of skewness; skewness.
schnelle Fouriertransformierte. fast Fourier transform.
Schrittweite. lag.
schwache Stationarität. weak stationarity.
Schwarzsche Ungleichung. Schwarz inequality.
Schwelle. sill.
selbstähnlich. selfsimilar.
selektiv. Selective.
Selektivitätskurve. selectivity curve.
semimarkovsche Zufallsfunktion. semimarkovian random function.
Semivariogramm. demivariogram; semivariogram.
Semivariogrammodell. semivariogram model.
Semivariogrammodell des reinen Nuggeteffekts. pure nugget effect semivariogram model.
Semivariogrammodell mit Locheffeekt. hole effect semivariogram model.
sequentielle Indikatorsimulation. sequential indicator simulation.
serielle Variationsfunktion. serial variation function.
Sichelsche t-Verteilung. Sichel t distribution.
Sigmafeld. sigma-field.
Simulation. simulation.
singuläre Matrix. singular matrix.
Skalarprodukt. scalar product.
Spalte einer Matrix. column of a matrix.
Spalte in einer Determinante. column of a determinant.
Spearmanscher Korrelationskoeffizient. Spearman correlation coefficient.
sphärisches Semivariogramm. spherical semivari-

ogram.
Spline. spline function.
Standardabweichung. standard deviation.
Standardfehler. standard error.
standardisierte Normalverteilung. standard normal distribution.
standardisierte Punktwerte. standard score; z-score.
stationär. stationary.
stationär endlicher Ordnung. stationary of finite order.
stationär im weiteren Sinne. stationary in the wide sense.
Stationarität. stationarity.
Statistik. statistic.
statistische Abhängigkeit. statistical dependence.
statistische Unabhängigkeit. statistical independence.
steilgipflig. leptokurtic.
stetige Zufallsvariable. continuous random variable.
Stichprobe. sample (a).
Stichprobenfunktion. sample function.
Stichprobenraum. sample description space; sample space.
Stichprobenumfang. sample size.
stochastische Abhängigkeit. stochastic dependence.
stochastischer Prozeß. stochastic process.
stochastische Unabhängigkeit. stochastic independence.
stochastische Variable. stochastic variable.
streng bedingt positiv definite Funktion. strictly conditionally positive definite function.
strenge Stationarität. strict stationarity; strong stationarity.
streng positiv definite Funktion. strictly positive definite function.
Streudiagramm. scattergram; scatterplot.
Struktur. structure.
Strukturanalyse. structural analysis.
strukturelle Funktion. structural function.
Strukturfunktion. structure function.
Strukturierung hoher Gehalte. high-grade structuration.
Stufenfunktion. step function.
Stützung einer regionalisierbaren Variablen. support (a).
Stützungseffekt. support effect.
Summenhäufigkeit. cumulative frequency.
symmetrische Matrix. symmetric matrix.
theoretisches Semivariogramm. theoretical semivariogram.
Theorie. theory.
Thiessen-Polygon. Thiessen polygon.
topologisches Wahrscheinlichkeitsmodell. topoprobabilistic model.
Träger einer Funktion. support (b).
Transferfunktion. transfer function.
Transformation. transformation.
Transformation in eine Gleichverteilung. uniform transformation.
Transformierte. transform.
transitiv. transitive.
transitive Methoden. transitive method.
transitives Kovariogram. transitive covariogram.
transitives Krigen. transitive kriging.
Translation zulässiger Koeffizienten. translation of permissible coefficients.
Transponierte einer Matrix. transpose of a matrix.
Trend. trend.
Übergangswahrscheinlichkeit. transition probability.
Überschätzung. overestimation.
Umgebung. neighborhood.
unbegrenzte Teilbarkeit. infinite divisibility.
unbegrenzte Zerlegbarkeit einer zufälligen Vereinigungsmenge. infinite divisibility of a union.
unempfindlich. resistant.
Ungewißheit. uncertainty.
unimodale Verteilung. unimodal distribution.
Universalitätsbedingung. universal condition.
Universalkrigen. universal kriging.
universell. universal.
unteres Quartil. lower quartile.
Unterschätzung. underestimation.
unverzerrt. unbiased.
Variable. variate.
Varianz. variance.
Varianzkorrekturfaktor. variance adjustment factor.
Variationskoeffizient. coefficient of variation.
Variogramm. variogram.
Variogramm der Ordnung α. variogram of order α.
Variographie. variography.
Vektor. vector.
Vektorraum. vector space.
verallgemeinerte Inkremente der Ordnung ν. generalized increments of order ν.
verallgemeinerte Kovarianzen der Ordnung ν. generalized covariance of order ν.
verallgemeinertes Semivariogramm. generalized semivariogram.
Verdünnungs-Zufallsvariable. dilution random function.
vereinbarungsgemäßer Parameter. conventional parameter.
Vereinigungsstabilität. union stability.
vergleichmäßigtes Semivariogramm. regularized semivariogram.
Vergleichmäßigung. regularization.
Verlustfunktion. loss function.
Versuch. experiment; trial.
Verteilung. distribution.
Verteilungsfunktion. cumulative distribution function; distribution function.
Vertrauensgrenze. confidence limit.
Vertrauensintervall. confidence interval.

Verzerrung. bias.
vollständiger Raum. complete space.
vollständiger Satz orthogonaler Funktionen. complete set of orthogonal functions.
vollständiger Satz orthogonaler Polynome. complete set of orthogonal polynomials.
Vorhersage. prediction.
Voronoi-Parkettierung. Voronoi tesselation.
Voronoi-Polygon. Voronoi polygon.
Voronoi-Zerlegung. Voronoi division.
Wahrscheinlichkeit. probability.
Wahrscheinlichkeitsdichte. probability density function.
Wahrscheinlichkeitsfunktion. probability function.
Wahrscheinlichkeitsintervall. probability interval.
Wahrscheinlichkeitskrigen. probability kriging.
Wahrscheinlichkeitsmaß. probability measure.
Wahrscheinlichkeitsraum. probability space.
Wahrscheinlichkeitsverteilung. probability distribution.
weiche Daten. soft data.
Wiener-Filter. Wiener filter.
Wiener–Levy-Prozeß. Wiener–Levy process.
Wiener-Prozeß. Wiener process.
weißes Rauschen. white noise.
Wigner–Seitz-Zelle. Wigner–Seitz cell.
Wölbung. coefficient of kurtosis; kurtosis.
Zeile einer Matrix. row of a matrix.
Zeile in einer Determinante. row of a determinant.
Zeitreihe. time series.
Zentil. centile.
zentraler Grenzwertsatz. central limit theorem.
Zentralwert. median.
Zentralwertindikatorkrigen. median indicator kriging.
zentrierte Kovarianz. centered covariance.
Zerlegung einer Menge. partition of a set.
Zerlegung in untere und obere Dreiecksmatrix. lower-upper (LU) decomposition.
Zielfunktion. objective function.
zonale Anisotropie. stratified anisotropy; zonal anisotropy.
Zufallsfeld. random field.
Zufallsfeld 2. Ordnung. second order random field.
Zufallsfunktion. random function.
Zufallsfunktion des weißen Rauschens. white random function.
Zufallskrigen. random kriging.
Zufallsmenge. random set.
Zufallsprozeß. random process.
Zufallsvariable. aleatory variable; random variable.
Zufallsvariable 2. Ordnung. second order random variable.
Zufallsversuch. random experiment.
zugrundeliegendes Semivariogramm. underlying semivariogram.
zulässige Koeffizienten der Ordnung ν. authorized coefficients of order ν; permissible coefficients of order ν.
zulässige Kombinationen der Ordnung ν. authorized combinations of order ν; permissible combinations of order ν.
Zulässigkeitskriterium. permissibility criteria.
zweidimensional. bivariate.
zweidimensionale Normalverteilung. bivariate normal distribution.
zweidimensionales Gaußsches Modell. bigaussian model.
zweidimensionale Verteilung. bivariate distribution.
zweiparametrische lognormale Verteilung. two-parameter lognormal distribution.

GREEK to ENGLISH

h-διασπορόγραμμα. h-scattergram.
jackknife. jackknife.
Sichel t κατανομή. Sichel t distribution.
αβεβαιότης. uncertainty.
αθροιστική συνάρτηση κατανομής. cumulative distribution function.
αθροιστική συχνότης. cumulative frequency.
ακολουθία Cauchy. Cauchy sequence.
ακολουθιακή προσομοίωση δείκτη. sequential indicator simulation.
ακρίβεια. accuracy.
ακριβής παρεμβολή. exact interpolator.
αμερόληπτος. unbiased.
άμεσο πρόβλημα. direct problem.
αμφιμονότιμη συνάρτηση. bijection.
αναγωγικός. recursion.
ανακτώμενα αποθέματα. recoverable reserves.
αναλογική επίδραση. proportional effect.
ανάλυση Fourier. Fourier analysis.
ανάλυση παραγόντων. factor analysis.
αναμενόμενη τιμή. expected value.
αναμόρφωση. anamorphosis; normal score transform.
αναμόρφωση Gauss. Gaussian anamorphosis.
αναπαράσταση μιάς εγγενούς τυχαίας συνάρτησης. representation of an intrinsic random function.
ανάστροφος ενός πίνακα. transpose of a matrix.
ανθεκτικός. resistant.
ανισότης Bessel. Bessel inequality.
ανισότης Buniakovski. Buniakovski inequality.
ανισότης Cauchy–Schwarz. Cauchy–Schwarz inequality.
ανισότης Schwarz. Schwarz inequality.
ανισοτροπία. anisotropy.
ανοικτός. open.
αντεπικύρωση. cross-validation.
αντικειμενική έννοια. objective meaning.
αντίστροφο πρόβλημα. inverse problem.
αντίστροφος ενός πίνακα. inverse of a matrix.
άνω τεταρτημόριο. upper quartile.
άξων. pivot.
άπειρος διαιρετότης. infinite divisibility.
άπειρος διαιρετότης μιάς ένωσης. infinite divisibility of a union.
απλή επίδραση κόκκου. pure nugget effect.
απλό κρίκιγκ. simple kriging.
αποδόμηση υψηλών περιεκτικοτήτων. high-grade destructuration.
απόκλιση τεταρτημορίου. quartile deviation.
από κοινού αντεπικύρωση. joint cross-validation.
από κοινού κατανομή πιθανότητας. joint probability distribution.
από κοινού ορθογωνικές τυχαίες διαδικασίες. jointly orthogonal random process.
από κοινού ορθογωνικές τυχαίες συναρτήσεις. jointly orthogonal random function.
από κοινού παραγοντικές τυχαίες συναρτήσεις. jointly factorable random function.
από κοινού πυκνότης πιθανότητας. joint probability density function.
απόλυτη διασπορά. absolute dispersion.
απομονωμένη τιμή. outlier.
αποσυνέλιξη. deconvolution.
αποσυσσώρευση. declustering.
αποτέλεσμα z. z-score.
αριθμητική μέση τιμή. arithmetic average.
αριθμητικός μέσος όρος. arithmetic mean.
αρμονική μέση τιμη. harmonic mean.
αρμονικός. harmonic.
αρνητική διωνυμική κατανομή. negative binomial distribution.
αρχή του τοπικού γνωρίσματος. regional characteristic principle.
ασθενής στασιμότης. weak stationarity.
αυστηρά γνησίως θετική συνάρτηση. strictly positive definite function.
αυστηρά δεσμευμένη γνησίως θετική συνάρτηση. strictly conditionally positive definite function.
αυστηρή στασιμότης. strict stationarity.
αυτοκρίκιγκ. autokrigeability; autokriging.
αυτο-ομοθετικός. selfsimilar.
αυτοσυνδιασπορά. autocovariance.
αυτοσυσχέτιση. autocorrelation.
αυτοσυσχετόγραμμα. autocorrelogram.
βαθμωτό ημιβαριόγραμμα. graded semivariogram.
βαριόγραμμα. variogram.
βαριόγραμμα βαθμού διάταξης. rank order variogram.
βαριόγραμμα τάξης α. variogram of order α.
βαριογραφία. variography.
βέλτιστο υπόλοιπο. optimal residual.
βηματική συνάρτηση. step function.
γειτονιά. neighborhood.
γειτονιά εργασίας. working neighborhood.
γειτονιά κρίκιγκ. kriging neighborhood.
γενικευμένες αυξήσεις τάξης ν. generalized increments of order ν.
γενικευμένη συνδιασπορά τάξης ν. generalized covariance of order ν.

γενικευμένο γραμμικό ημιβαριόγραμμα. generalized semivariogram.
γενικό σχετικό ημιβαριόγραμμα. general relative semivariogram.
γεωμετρική ανισοτροπία. geometric anisotropy.
γεωμετρική μέση τιμή. geometric mean.
γεωμετρική υποστήριξη. geometric support.
γεωμετρικό πεδίο. geometric field.
γεωμετρικός μέσος όρος. geometric average.
γεωμετρικό συνδιασπορόγραμμα. geometric covariogram.
γεωστατιστική. geostatistics.
γινόμενο αριθμού επι πίνακα. product of a scalar and a matrix.
γινόμενο πινάκων. product of matrices.
γνησίως θετική. positive definite.
γραμμή ενός πίνακα. row of a matrix.
γραμμή μιάς ορίζουσας. row of a determinant.
γραμμή παλινδρόμησης. regression line.
γραμμική ανεξαρτησία. linear independence.
γραμμική γεωστατιστική. linear geostatistics.
γραμμική εξάρτηση. linear dependence.
γραμμική πολλαπλότης. linear manifold.
γραμμικό ημιβαριόγραμμα. linear semivariogram.
γραμμικό μοντέλο συσχέτισης. linear model of coregionalization.
γραμμικός συνδιασμός μεταβλητών. linear combination of variables.
γραμμικος χώρος. linear space.
γραμμικό φιλτράρισμα. linear filtering.
δεδομένα. data.
δείγμα. sample.
δειγματοσυνάρτηση. realization; sample function.
δειγματοχώρος. sample description space.
δείκτης. indicator.
δεκατημόριο. decile.
δέλτα Dirac. Dirac delta function.
δέλτα Kronecker. Kronecker delta.
δεσμευμένη αμεροληψία. conditional unbiasedness.
δεσμευμένη κατανομή. conditional distribution.
δεσμευμένη μεροληψία. conditional bias.
δεσμευμένη μέση τιμή. conditional expectation.
δεσμευμένη μη-μεροληψία. conditional nonbias.
δεσμευμένη πιθανότης. conditional probability.
δεσμευμένη προσομοίωση. coconditional simulation; conditional simulation.
δεσμευμένη συνάρτηση πυκνότητος πιθανότητος. conditional probability density function.
δεσμευμένη συνπροσομοίωση. conditional cosimulation.
δεσμευμένος θετικά ορισμένος ν τάξης. conditionally positive definite of order ν.
διαβάθμιση. grading.
διάγραμμα ρόδου. rose diagram.
διάγραμμα συχνότητος. frequency diagram.
διαγώνιος διμεταβλητή συνάρτηση πυκνότητος. diagonal bivariate density function.
διαγώνιος πίνακας. diagonal matrix.
διαγωνοποιημένο μητρώο. diagonalized matrix.
διαδικασία Brown. Brownian motion process.
διαδικασία Markov. Markov process.
διαδικασία Poisson. Poisson process.
διαδικασία Wiener. Wiener process.
διαδικασία Wiener–Levy. Wiener–Levy process.
διαζευτικό κρίκιγκ. disjunctive kriging.
διαίρεση Voronoi. Voronoi division.
διακεκριμμένη τυχαία μεταβλητή. discrete random variable.
διακεκριμμένο μοντέλο Gauss. discrete Gaussian model.
διαμερισμός ενος συνόλου. partition of a set.
διάμεσος. median.
διάνυσμα. vector.
διανυσματικός χώρος. vector space.
διασπορά. variance.
διασπορά εκτίμησης. estimation variance.
διασπορά επέκτασης. extension variance.
διασπορά κρίκιγκ. kriging variance.
διασπορά σκέδασης. dispersion variance.
διασπορόγραμμα. scattergram; scatterplot.
διάσταση Hausdorff–Besicovitch. Hausdorff–Besicovitch dimension.
διάσταση ομοιότητος. similarity dimension.
διάστημα ενδοτεταρτημορίου. interquartile range.
διάστημα εμπιστοσύνης. confidence interval.
διάστημα ολοκλήρωσης. integral range.
διάστημα περιορισμού. constraint interval.
διάστημα πιθανότητας. probability interval.
διασυσχέτιση. cross-correlation.
διατύπωση μεγίστης εντροπίας. maximum entropy formalism.
δικανονική κατανομή. bigaussian model.
δίκορφη κατανομή. bimodal distribution.
διμεταβλητή. bivariate.
διμεταβλητή κανονική κατανομή. bivariate normal distribution.
διμεταβλητή κατανομή. bivariate distribution.
διωνυμική κατανομή. binomial distribution.
δοκιμή. trial.
δομή. structure.
δομική ανάλυση. structural analysis.
δομική συνάρτηση. structural function; structure function.
δυϊκό κρίκιγκ. dual kriging.
δυναμοημιβαριόγραμμα. power semivariogram model.
εγγενές. intrinsic.
εγγενές κρίκιγκ. intrinsic kriging.
εγγενές χαρακτηριστικό. intrinsic attribute.
εγγενής διασπορά. intrinsic dispersion.
εγγενής συνάρτηση. intrinsic function.
εγγενής συσχέτιση. intrinsic correlation.

εγγενής τυχαία συνάρτηση τάξης ν. intrinsic random function of order ν.
εγγενής υπόθεση. intrinsic hypothesis.
εγγύς πολύγωνο. proximal polygon.
εγκάρσια συνδιασπορά. cross-covariance.
εγκάρσια συνδιασπορά δείκτη. indicator cross-covariance.
εγκάρσια συσχέτιση. cross-correlation.
εγκάρσιο βαριόγραμμα. cross-variogram.
εγκάρσιο διάγραμμα. cross-plot.
εγκάρσιο ημιβαριόγραμμα. cross-semivariogram.
εγκλωβιστική κατασκευή. nested structure.
εγκλωβιστικό ημιβαριόγραμμα. nested semivariogram.
εκατοστημόριο. centile; percentile.
έκβαση. outcome.
εκθετική κατανομή. exponential distribution.
εκθετικό ημιβαριόγραμμα. exponential semivariogram.
εκλεκτικός. selective.
εκτίμηση. estimate; estimation.
εκτιμητής. estimator.
εκτιμητής ελαχίστου μετρικού λάθους. minimum error norm estimator.
εκτιμητής μεγίστης πιθανοφάνειας. maximum likelihood estimator.
εκτιμητής τύπου E. E-type estimate.
ελάχιστο μέσο τετραγωνικό λάθος εκτίμησης. minimum mean square error estimator.
ελλειπτική ανισοτροπία. elliptical anisotropy.
έλλειψη Krige. Krige's ellipse.
εμβέλεια. range.
έμμεση λογοκανονική διόρθωση. indirect lognormal correction.
ενδεχόμενο. event.
εντροπία. entropy.
ενυπάρχον ημιβαριόγραμμα. underlying semivariogram.
εξίσωση Parseval. Parseval's equation.
εξομάλυνση. smoothing.
επακρίβεια. precision.
επέκταση. extrapolation.
επίδραση κόκκου. nugget effect.
επίδραση οθόνης. screen effect.
επίδραση οπής. hole effect.
επίδραση περιορισμού. constraint effect.
επίδραση πληροφόρησης. information effect.
επίδραση φορέα. support effect.
επικύρωση. validation.
επίπεδο αναφοράς. datum.
επιτόπου αποθέματα. reserves in place.
επιτρεπτοί συνδιασμοί τάξης ν. authorized combinations of order ν; permissible combinations of order ν.
επιτρεπτοί συντελεστές τάξης ν. authorized coefficients of order ν; permissible coefficients of order ν.
επιχειρησιακή έννοια. operational concept.
εργοδική ιδιότης. ergodic property.
εργοδικό θεώρημα. ergodic theorem.
εργοδικότης. ergodicity.
εσχάρα. grid.
εσωδικτυακή κατασκευή. intermeshed structure.
εσωτερικό γινόμενο. inner product; scalar product.
εσωτερικός. interior.
ετεροσκεδαστικότης. heteroscedastic.
ετεροτοπικός. heterotopic.
Ευκλίδειος χώρος. euclidean space.
εύρωστος. robust.
ευστάθεια ένωσης. union stability.
ζώνη επιρροής. zone of influence.
ζωνική ανισοτροπία. zonal anisotropy.
ημιβαριόγραμμα. demivariogram; semivariogram.
ημιβαριόγραμμα de Wijs. de Wijsian semivariogram.
ημιβαριόγραμμα Gauss. Gaussian semivariogram.
ημιβαριόγραμμα απλής επίδρασης κόκκου. pure nugget effect semivariogram model.
ημιβαριόγραμμα δύναμης. power semivariogram model.
ημιβαριόγραμμα επίδρασης οπής. hole effect semivariogram model.
θαμπά δεικτικά δεδομένα. fuzzy indicator data.
θεώρημα Bayes. Bayes' theorem.
θεωρητικό ημιβαριόγραμμα. theoretical semivariogram.
θεωρία. theory.
ιδιάζων πίνακας. singular matrix.
ιδιοτιμή. eigenvalue.
ιδιοτιμή ενός πίνακα eigenvalue of a matrix.
ισογραμμή. isoline.
ισοπαραγοντική διμεταβλητή συνάρτηση πυκνότητας. isofactorial bivariate density function.
ισοπληθής. isopleth.
ισοτοπικό. isotopic.
ισοτροπία. isotropy.
ισοϋψής. contour.
ιστόγραμμα. histogram.
ισχυρή στασιμότης. strong stationarity.
καθολική συνθήκη. universal condition.
καθολικό κρίκιγκ. universal kriging.
καθολικός. universal.
καμπύλη εκλεκτικότητος. selectivity curve.
καμπύλη συχνότητος. frequency curve.
κανονικές εξισώσεις. normal equations.
κανονική κατανομή. normal distribution.
κανονικό πρόβλημα. normal problem.
κανονικό σύστημα. regular system.
κατά ζεύγη σχετικό ημιβαριόγραμμα. pairwise relative semivariogram.
κατανομή. distribution.
κατανομή Beta. beta distribution.
κατανομή Gauss. Gaussian distribution.
κατανομή Poisson. Poisson distribution.
κατανομή γάμμα. gamma distribution.

κατανομή πιθανότητας. probability distribution.
κατασκευή. structure
κατώτατο όριο. infimum.
κατώτερη-ανώτερη αποσύνθεση (ανάλυση). lower-upper (LU) decomposition.
κάτω τεταρτημόριο. lower quartile.
κατώφλι. sill.
κελί Wigner–Seitz. Wigner–Seitz cell.
κελί Meijering. Meijering's cell.
κεντρική συνδιασπορά. centered covariance.
κεντρικό οριακό θεώρημα. central limit theorem.
κινητός γραμμικός μέσος. moving linear average.
κλασματική διάσταση. fractal dimension.
κλειστό. closed.
κλίμακα εργασίας. working scale.
κοκρίκιγκ. cokriging.
κοκρίκιγκ δείκτη. indicator cokriging.
κορυφή. mode.
κρίκιγκ. kriging.
κρίκιγκ δείκτη. indicator kriging.
κρίκιγκ δείκτη κυρίων συνιστωσών. indicator principal component kriging.
κρίκιγκ διαμέσου ενός δείκτη. median indicator kriging.
κρίκιγκ πολλαπλού δείκτη. multiple indicator kriging.
κρίκιγκ πολυκανονικό. multigaussian kriging.
κρίκιγκ υπολοίπου. residual kriging.
κριτήρια επιτρεπτότητας. permissibility criteria.
κριτήριο αντικειμενικότητας. objectivity criterion.
κυβικό ημιβαριόγραμμα. cubic semivariogram.
κύμανση. range.
κύρια διαγώνιος ενός πίνακα. main diagonal of a matrix; principal diagonal of a matrix.
κύρια συνιστώσα. principal component.
κυρτή συνάρτηση. convex function.
κύρτωση. kurtosis.
λάθος. error.
λεπτοκυρτικός. leptokurtic.
λευκή τυχαία συνάρτηση. white random function.
λευκός θόρυβος. white noise.
λογαριθμικό ημιβαριόγραμμα. logarithmic semivariogram.
λογαριθμικό κρίκιγκ. logarithmic kriging.
λογική έκφραση. rational expression.
λογοκανονική διόρθωση. lognormal correction.
λογοκανονική κατανομή. lognormal distribution.
λογοκανονική κατανομή δύο παραμέτρων. two-parameter lognormal distribution.
λογοκανονική κατανομή τριών παραμέτρων. three-parameter lognormal distribution.
λογοκανονικό κρίκιγκ. lognormal kriging.
λοξότης. skewness.
μαλακά δεδομένα. soft data.
μαντόγραμμα. madogram.
μέγεθος δείγματος. sample size.
μέθοδος Choleski. Choleski's method.
μέθοδος Monte Carlo. Monte Carlo method.
μέθοδος εξωτερικής τάσης. external drift method.
μέθοδος στρεφόμενης δέσμης. turning bands method.
μέθοδος τυχαίου κέρματος. random coin method; random tokens method.
μέθοδος των ελαχίστων τετραγώνων. least squares method.
μέθοδος των πολλαπλασιαστών Lagrange. Lagrange's method of multipliers.
μεροληψία. bias.
μέση τιμή. average value; mean.
μέσο ημιβαριόγραμμα. average semivariogram; mean semivariogram.
μεσοκυρτικός. mesokurtic.
μέσος όρος. average.
μέσο τετραγωνικό λάθος. mean square error.
μεταβατικές μέθοδοι. transitive method.
μεταβατική πιθανότης. transition probability.
μεταβατικό κρίκιγκ. transitive kriging.
μεταβατικός transitive.
μεταβατικό συνδιασπορόγραμμα. transitive covariogram.
μεταγενέστερη κατανομή. posterior distribution.
μετασχηματίζω. transform.
μετασχηματισμός. transformation.
μετασχηματισμός Fourier. Fourier transform.
μετασχηματισμός Laplace. Laplace transform.
μετασχηματισμός βαθμού διάταξης. rank order transform.
μεταφορά επιτρεπτών συντελεστών. translation of permissible coefficients.
μετρικός χώρος. metric space.
μέτρο. norm; measure.
μέτρο Hausdorff. Hausdorff measure.
μέτρο πιθανότητας. probability measure.
μη γραμμική γεωστατιστική. nonlinear geostatistics.
μη εργοδική πειραματική συνδιασπορά. nonergodic experimental covariance.
μη ιδιάζον σύστημα. nonsingular system.
μη κεντρική συνδιασπορά. noncentered covariance.
μη παραμετρική γεωστατιστική. nonparametric geostatistics.
μητρώο. matrix.
μητρώο Hermite. Hermitian matrix.
μικροεργοδικότης. microergodicity.
μοναδικό κρίκιγκ δείκτη. single indicator kriging.
μονόκορφη κατανομή. unimodal distribution.
μονομεταβλητή. univariate.
μονοσκοπική. monoscopic.
μοντέλο. model.
μοντέλο Hermite. Hermite model.
μοντέλο αλλαγής φορέα. change of support model.
μοντέλο ημιβαριογράμματος. semivariogram model.
μπλοκ κρίκιγκ. block kriging.
μωσαϊκό Dirichlet. Dirichlet tesselation.

μωσαϊκό Voronoi. Voronoi tesselation.
μωσαϊκό μοντέλο. mosaic model.
νόμος των μεγάλων αριθμών. law of large numbers.
νόμος χωρικού μεγέθους. regional magnitude principle.
ολική εκτίμηση. global estimation.
ολικός. global.
ολισθαίνουσα αναπαράσταση. sliding representation.
ολισθαίνουσα γειτονιά. sliding neighborhood.
ολοκληρώσιμη συνάρτηση. integrable function.
ολοκληρώσιμη συνάρτηση ως προς μια δεύτερη συνάρτηση. integrable function with respect to a second function.
ολότητα. universe.
ομαλοποιημένο ημιβαριόγραμμα. regularized semivariogram.
ομαλοποίηση. regularization.
ομοιογένεια. homogeneity.
ομοιοδιάστατος. codimension.
ομοιόμορφη δέσμευση. uniform conditioning.
ομοιόμορφη κατανομή. uniform distribution.
ομοιόμορφος μετασχηματισμός. uniform transformation.
ομοσκεδαστικότης. homoscedastic.
ορθογωνικά πολυώνυμα. orthogonal polynomial.
ορθογωνική συνάρτηση. orthogonal function.
ορθογωνική τυχαία διαδικασία. orthogonal random process.
ορθογωνική τυχαία συνάρτηση. orthogonal random function.
ορθογωνικό διάνυσμα. orthogonal vector.
ορθοκανονική συνάρτηση. orthonormal function.
ορθοκανονικό διάνυσμα. orthonormal vector.
ορίζουσα. determinant.
ορίζουσα ενός πίνακα. determinant of a matrix.
όριο εμπιστοσύνης. confidence limit.
παλινδρόμηση. regression.
παλμός. impulse.
πανσκοπική. panscopic.
παραβολική συμπεριφορά. parabolic behavior.
παραβολικό ημιβαριόγραμμα. parabolic semivariogram.
παραγοντική τυχαία διαδικασία. factorable random process.
παραγοντική τυχαία συνάρτηση. factorable random function.
παραγοντικό κρίκιγκ. factorial kriging.
παράγων προσαρμογής διασποράς. variance adjustment factor.
παραδοσιακή παράμετρος. conventional parameter.
παράθυρο. window.
παρεμβολή. interpolation.
πεδίο Borel. Borel field.
πεδίο ορισμού. domain; field.
πεδίο σίγμα. sigma-field.
πείραμα. experiment.
πείραμα Bernoulli. Bernoulli experiment.
πειραματική συνδιασπορά. experimental covariance.
πειραματικό ημιβαριόγραμμα. experimental semivariogram.
πεντασφαιρικό ημιβαριόγραμμα. penta-spherical semivariogram.
περιθωριακή κατανομή. marginal distribution.
περιθωριακή πυκνότης πιθανότητος. marginal probability density function.
περιορισμένο υπόλοιπο. reduced residual.
πιθανότης. probability.
πιθανοτικό κρίκιγκ. probability kriging.
πιθανοτικό χωρικό ημιβαριόγραμμα. probabilistic regional semivariogram.
πίνακας. matrix.
πίνακας συνδιασποράς. covariance matrix.
πλατοκυρτικός. platykurtic.
πληθυσμός. population; universe.
πλήρες σύνολο ορθογωνικών πολυωνύμων. complete set of orthogonal polynomials.
πλήρες σύνολο ορθογωνικών συναρτήσεων. complete set of orthogonal functions.
πλήρης χώρος. complete space.
πολλαπλασιαστής Lagrangre. Lagrange multiplier.
πολλαπλότης. manifold.
πολύγωνο Thiessen. Thiessen polygon.
πολύγωνο Voronoi. Voronoi polygon.
πολύγωνο επιρροής. polygon of influence.
πολυκανονική κατανομή. multinormal distribution.
πολύκορφη κατανομή. multimodal distribution.
πολυμεταβλητή. multivariate.
πολυμεταβλητή κατανομή. multivariate distribution.
πολυσκοπικός. polyscopic.
πολυώνυμα Hermite. Hermite polynomials.
πολυώνυμα Laguerre. Laguerre polynomials.
πολυώνυμο *n* βαθμού. polynomial of degree *n*.
πολυώνυμο διαφόρων μεταβλητών. polynomial in several variables.
ποσοστημόριο. quantile.
ποσοστιαίο σημείο. fractile.
πρόβλεψη. prediction.
προβολικό θεώρημα. projection theorem.
προδρομικό πρόβλημα. forward problem.
προμεταγενέστερη κατανομή. pre-posterior distribution.
προσαρμογή ημιβαριογράμματος. semivariogram fitting.
προσαρμογή καμπύλης. curve fitting.
προσδιορισμός υψηλών περιεκτικοτήτων. high grade structuration.
προσδιοριστική γεωστατιστική. deterministic geostatistics.
προσθετική συνολο-συνάρτηση. additive set function.
προσθετικό θεώρημα. additivity theorem.
προσομοίωση. simulation.

προτέρα διασπορά. a priori variance.
προτέρα κατανομή. prior distribution.
Πυθαγόρειο θεώρημα. Pythagorean theorem.
ροδόγραμμα. rodogram.
ροπή. moment.
ροπή μιάς κατανομής. moment of a distribution.
σ–άλγεβρα. σ–algebra.
σημειακό κρίκιγκ. point kriging; punctual kriging.
σημείο συσσώρευσης. accumulation point.
σκληρά δεδομένα. hard data.
στάσιμος. stationary.
στάσιμος με την ευρεία έννοια. stationary in the wide sense.
στάσιμος πεπερασμένης τάξης. stationary of finite order.
στασιμότης. stationarity.
στατιστική. statistic.
στατιστική ανεξαρτησία. statistical independence.
στατιστική εξάρτηση. statistical dependence.
στατιστική μεγίστης πιθανοφάνειας. maximum likelihood statistic.
στήλη δεδομένων δείκτη. indicator data column.
στήλη ενός πίνακα. column of a matrix.
στήλη μιάς ορίζουσας. column of a determinant.
στοιχείο ενός πίνακα. element of a matrix.
στοιχείο μιας ορίζουσας. element of a determinant.
στοχαστική ανεξαρτησία. stochastic independence.
στοχαστική διαδικασία. stochastic process.
στοχαστική εξάρτηση. stochastic dependence.
στοχαστική μεταβλητή. stochastic variable.
στρωματοποιημένη ανισοτροπία. stratified anisotropy.
στρωματοποιημένη τυχαία δειγματοληψία. stratified random sampling.
συζυγής μιγαδικός αριθμός. conjugate complex number.
συμβάν. event.
συμμετρικός πίνακας. symmetric matrix.
συμπαγής. compact.
συνάρτηση. function.
συνάρτηση Beta. beta function.
συνάρτηση F. F-function.
συνάρτηση Heaviside. Heaviside's function.
συνάρτηση ανάκτησης. recovery function.
συνάρτηση αντικείμενο. objective function.
συνάρτηση απώλειας. loss function.
συνάρτηση αυτοσυσχέτισης. autocorrelation function.
συνάρτηση βάσης. basis function.
συνάρτηση γάμμα. gamma function.
συνάρτηση κατανομής. distribution function.
συνάρτηση μεταφοράς. transfer function.
συνάρτηση πιθανότητας. probability function.
συνάρτηση πιθανοφάνειας. likelihood function.
συνάρτηση πυκνότητος πιθανότητος. probability density function.
συνάρτηση σπλίνας. spline function.
συνάρτηση συχνότητας. frequency function.
συνδιασπορά. covariance.
συνδιασπορά δείκτη. indicator covariance.
συνέλιξη. convolution.
συνέπεια. consistency.
συνεχής τυχαία μεταβλητή. continuous random variable.
συνεχόμενη συνάρτηση μεταβολής. serial variation function.
σύνηθες κρίκιγκ. ordinary kriging.
σύνθεση Fourier. Fourier synthesis.
σύνολο. ensemble.
συνπροσομοίωση. cosimulation.
συντελεστής αυτοσυσχέτισης. autocorrelation coefficient.
συντελεστής εγκάρσιας συσχέτισης. cross-correlation coefficient.
συντελεστής ενός στοιχείου μιάς ορίζουσας. cofactor of an element in a determinant.
συντελεστής κύρτωσης. coefficient of kurtosis.
συντελεστής λοξότητας. coefficient of skewness.
συντελεστής μεταβλητότητας. coefficient of variation.
συντελεστής συσχέτισης. correlation coefficient.
συντελεστής συσχέτισης Pearson. Pearson correlation coefficient.
συντελεστής συσχέτισης Spearman. Spearman correlation coefficient.
συντελεστής υπέρβασης. coefficient of excess.
συσσώρευση. accumulation.
συσχέτιση. correlation.
συσχετισμένη διόρθωση. affine correction.
συσχετισμένος μετασχηματισμός. affinity transformation.
συσχετισμός. coregionalization.
συσχετόγραμμα. correlogram; crosscorrelogram.
σφαιρικό ημιβαριόγραμμα. spherical semivariogram.
σχέδιο ποσοστημόριου. q-q plot.
σχεδόν-στάσιμος. quasi-stationary.
σχέση Cartier. Cartier's relation.
σχετική επίδραση κόκκου. relative nugget effect.
σχετικό ημιβαριόγραμμα. relative semivariogram.
σχηματισμός. configuration.
τάξη ορίζουσας. order of a determinant.
τάξη τετραγωνικού πίνακα. order of a square matrix.
τάση. drift; trend.
ταυτοτικός πίνακας. identity matrix.
ταχύς μετασχηματισμός Fourier. fast Fourier transform.
τελεστής. operator.
τεταρτημόριο. quartile.
τετραγωνικός πίνακας. square matrix.
τομικός. fractal.
τοπική εκτίμηση. local estimation.

τοπικό γνώρισμα. regional characteristic.
τοπικός. local.
τοπικό σχετικό ημιβαριόγραμμα. local relative semivariogram.
τοποπιθανοτικό μοντέλο. topoprobabilistic model.
τριγωνισμός Delaunay. Delaunay triangulation.
τυπική απόκλιση. standard deviation.
τυπική κανονική κατανομή. standard normal distribution.
τυπικό λάθος. standard error.
τυπικός βαθμός. standard score.
τύπος του de Wijs. de Wijs formula.
τύπος Krige. Krige's formula.
τύπος του Rodrigues. Rodrigues formula.
τυχαία διαδικασία. random process.
τυχαία μεταβλητή. aleatory variable; random variable; variate.
τυχαία συνάρτηση. random function.
τυχαία συνάρτηση Boole. Boolean random function.
τυχαία συνάρτηση αραίωσης. dilution random function.
τυχαία συνάρτηση δευτέρας τάξης. second order random variable.
τυχαία συνάρτηση ημι-Markov. semimarkovian random function.
τυχαίο κρίκιγκ. random kriging.
τυχαίο πεδίο. random field.
τυχαίο πεδίο δευτέρας τάξης. second order random field.
τυχαίο πείραμα. random experiment.
τυχαίο σύνολο. random set.
υπερεκτίμηση. overestimation.
υποδιαίρεση Delaunay. Delaunay subdivision.
υποεκτίμηση. underestimation.
υπόθεση διατήρησης του νόμου. conservation of law hypothesis.
υπόλοιπο. residual.
υποστήριξη. support.
υστέρηση. lag.
φαινομενική ασυνέχεια στήν αρχή. apparent discontinuity at the origin.
φιλτράρισμα. filtering.
φίλτρο Wiener. Wiener filter.
φορέας. support.
χαρακτηριστική εξίσωση ενός πίνακα. characteristic equation of a matrix.
χαρακτηριστική τιμή. eigenvalue.
χαρακτηριστικό διάνυσμα. eigenvector.
χρονοσειρά. time series.
χωρική ανεξαρτησία. spatial independence.
χωρική εξάρτηση. spatial dependence.
χωρική μεταβλητή. regionalized variable.
χωρικό ημιβαριόγραμμα. regional semivariogram.
χωρικό μέγεθος. regional magnitude
χωρικός μετασχηματισμός δευτέρου είδους. space transformation of the second kind.
χωρικός μετασχηματισμός πρώτου είδους. space transformation of the first kind.
χωρητικότης Choquet. Choquet's capacity.
χώρος Hilbert. Hilbert space.
χώρος δείγματος. sample space.
χώρος ενδεχομένων. event space.
χώρος εσωτερικού γινομένου. inner product space.
χώρος πιθανότητας. probability space.

PORTUGUESE to ENGLISH

aberto. open.
achatamento. kurtosis.
acontecimento. event.
acumulação. accumulation; grading.
ajustamento de curvas. curve fitting.
ajustamento do variograma. semivariogram fitting.
alcance. range.
alcance integral. integral range.
álgebra-σ. $\sigma-$algebra; sigma-field.
amostra. datum (b); sample.
amostragem aleatória estratificada. stratified random sampling.
amplitude. range.
análise de Fourier. Fourier analysis.
análise estrutural. structural analysis.
análise factorial. factor analysis.
anamorfose. anamorphosis.
anamorfose gaussiana. Gaussian anamorphosis; normal score transform.
anisotropia. anisotropy.
anisotropia elíptica. elliptical anisotropy.
anisotropia geométrica. geometric anisotropy.
anisotropia zonal. stratified anisotropy; zonal anisotropy.
assimetria. skewness.
atenuação. smoothing.
autocorrelação. autocorrelation.
autocorrelograma. autocorrelogram.
autocovariância. autocovariance.
autohomotético. selfsimilar.
autokrigabilidade. autokrigeability.
autokrigagem. autokriging.
bidimensional. bivariate.
bijecção. bijection.
campo. domain; field.
campo aleatório. random field.
campo aleatório de ordem 2. second order random field.
campo de Borel. Borel field.
campo geométrico. geometric field.
capacidade de Choquet. Choquet's capacity.
característica intrínseca. intrinsic attribute.
característica regional. regional characteristic.
célula de Meijering. Meijering's cell.
célula de Wigner–Seitz. Wigner–Seitz cell.
codimensão. codimension.
coeficiente de achatamento. coefficient of kurtosis.
coeficiente de assimetria. coefficient of skewness.
coeficiente de autocorrelação. autocorrelation coefficient.
coeficiente de correlação. correlation coefficient.
coeficiente de correlação cruzada. cross-correlation coefficient.
coeficiente de correlação de Pearson. Pearson correlation coefficient.
coeficiente de correlação de Spearman. Spearman correlation coefficient.
coeficiente de excesso. coefficient of excess.
coeficiente de variação. coefficient of variation.
coeficientes autorizados de ordem ν. authorized coefficients of order ν; permissible coefficients of order ν.
coerência. consistency.
cokrigagem. cokriging.
cokrigagem da indicatriz. indicator cokriging.
coluna de uma matriz. column of a matrix.
coluna de um determinante. column of a determinant.
coluna dos dados da indicatriz. indicator data column.
combinação linear de variáveis. linear combination of variables.
combinações autorizadas de ordem ν. authorized combinations of order ν; permissible combinations of order ν.
compacto. compact.
componente algébrico de um elemento num determinante. cofactor of an element in a determinant.
componente principal. principal component.
comportamento parabólico. parabolic behavior.
conceito de operacionalidade. operational concept.
condição de universalidade. universal condition.
condicionalmente definida positiva de ordem ν. conditionally positive definite of order ν.
condicionamento uniforme. uniform conditioning.
configuração. configuration.
conjunto. ensemble.
conjunto aleatório. random set.
contagem normalizada. standard score; z-score.
contorno. contour.
convolução. convolution.
correcção afim. affine correction.
correcção lognormal. lognormal correction.
correcção lognormal indirecta. indirect lognormal correction.
corregionalização. coregionalization.
correlação. correlation.
correlação cruzada. cross-correlation.
correlação intrínseca. intrinsic correlation.
correlograma. correlogram.
correlograma cruzado. crosscorrelogram.

cossimulação. cosimulation.
cossimulação condicional. coconditional simulation; conditional cosimulation.
covariância. covariance.
covariância centrada. centered covariance.
covariância cruzada. cross-covariance.
covariância cruzada da indicatriz. indicator cross-covariance.
covariância da indicatriz. indicator covariance.
covariância experimental. experimental covariance.
covariância experimental não ergódica. nonergodic experimental covariance.
covariância generalizada de ordem v. generalized covariance of order v.
covariância não centrada. noncentered covariance.
covariograma geométrico. geometric covariogram.
covariograma transitivo. transitive covariogram.
critério de objectividade. objectivity criterion.
critérios de coerência. permissibility criteria.
curtose. kurtosis.
curva de frequências. frequency curve.
curva de selectividade. selectivity curve.
dado. datum (b).
dados. data.
dados fiáveis. hard data.
dado soft. soft data.
dados vagos de indicatriz. fuzzy indicator data.
dado vago. soft data.
datum. datum (a).
decil. decile.
decomposição L.U. lower-upper (LU) decomposition.
definido positivo. positive definite.
dependência espacial. spatial dependence.
dependência estatística. statistical dependence.
dependência estocástica. stochastic dependence.
dependência linear. linear dependence.
deriva. drift.
desaglomeração. declustering.
descontinuidade aparente na origem. apparent discontinuity at the origin.
desconvolução. deconvolution.
desestruturação dos altos teores. high-grade destructuration.
desigualdade de Bessel. Bessel inequality.
desigualdade de Buniakovski. Buniakovski inequality.
desigualdade de Cauchy–Schwarz. Cauchy–Schwarz inequality.
desigualdade de Schwarz. Schwarz inequality.
desvio-padrão. standard deviation.
determinante. determinant.
determinante de uma matriz. determinant of a matrix.
diagonal principal de uma matriz. main diagonal of a matrix; principal diagonal of a matrix.
diagrama q-q. q-q plot.
diagrama de frequências. frequency diagram.
diagrama polar. rose diagram.
dimensão da amostra. sample size.
dimensão de Hausdorff–Besicovitch. Hausdorff–Besicovitch dimension.
dimensão de similaridade. similarity dimension.
dimensão fractal. fractal dimension
dispersão absoluta. absolute dispersion.
dispersão intrínseca. intrinsic dispersion.
distribuição. distribution.
distribuição à priori. prior distribution.
distribuição Beta. beta distribution.
distribuição bidimensional. bivariate distribution.
distribuição bimodal. bimodal distribution.
distribuição binomial. binomial distribution.
distribuição binomial negativa. negative binomial distribution.
distribuição condicionada. conditional distribution; posterior distribution.
distribuição de Poisson. Poisson distribution.
distribuição de probabilidade. probability distribution.
distribuição exponencial. exponential distribution.
distribuição Gama. gamma distribution.
distribuição gaussiana. Gaussian distribution.
distribuição lognormal. lognormal distribution.
distribuição lognormal a dois parâmetros. two-parameter lognormal distribution.
distribuição lognormal a três parâmetros. three-parameter lognormal distribution.
distribuição marginal. marginal distribution.
distribuição multidimensional. multivariate distribution.
distribuição normal. normal distribution.
distribuição normal bidimensional. bivariate normal distribution.
distribuição normal multidimensional. multinormal distribution.
distribuição normal normalizada. standard normal distribution.
distribuição normal reduzida. standard normal distribution.
distribuição plurimodal. multimodal distribution.
distribuição pré-posterior. pre-posterior distribution.
distribuição t de Sichel. Sichel t distribution.
distribuição uniforme. uniform distribution.
distribuição unimodal. unimodal distribution.
divisão de Voronoi. Voronoi division.
domínio. domain.
efeito de cabeça de cavalo. hole effect.
efeito de constrangimento geométrico. constraint effect.
efeito de écran. screen effect.
efeito de informação. information effect.
efeito de pepita. nugget effect; pure nugget effect.
efeito de pepita relativo. relative nugget effect.
efeito de suporte. support effect.
efeito proporcional. proportional effect.

elemento de uma matriz. element of a matrix.
elemento de um determinante. element of a determinant.
elipse de Krige. Krige's ellipse.
ensaio. trial.
entropia. entropy.
enviesamento. bias.
enviesamento condicional. conditional bias.
equação característica de uma matriz. characteristic equation of a matrix.
equação de Parseval. Parseval's equation.
equações normais. normal equations.
ergodicidade. ergodicity.
erro. error.
erro padrão. standard error.
erro quadrático médio. mean square error.
escala de trabalho. working scale.
espaço amostral. sample description space.
espaço completo. complete space.
espaço das amostras. sample space.
espaço de Hilbert. Hilbert space.
espaço de probabilidade. probability space.
espaço dos acontecimentos. event space.
espaço dos resultados. sample description space.
espaço euclidiano. euclidean space.
espaço métrico. metric space.
espaço vectorial. linear space; vector space.
espaço vectorial munido de produto interno. inner product space.
esperança matemática. expected value; mean.
esperança matemática condicionada. conditional expectation.
esquema booleano. Boolean random function.
esquema de de Wijs. de Wijsian semivariogram.
estabilidade para a união. union stability.
estacionariedade. stationarity.
estacionariedade estrita. strict stationarity.
estacionariedade forte. strong stationarity.
estacionariedade fraca. weak stationarity.
estacionário. stationary.
estacionário de ordem finita. stationary of finite order.
estacionário em sentido lato. stationary in the wide sense.
estatístico. statistic.
estimação. estimation.
estimação global. global estimation.
estimação local. local estimation.
estimador. estimate; estimator.
estimador de máxima verosimilhança. maximum likelihood estimator; maximum likelihood statistic.
estimador de tipo-E. E-type estimate.
estrutura. structure.
estruturação dos altos teores. high-grade structuration.
estrutura embricada. intermeshed structure; nested structure.
exactidão. accuracy.
experiência. experiment.
experiência aleatória. random experiment.
expressão racional. rational expression.
extrapolação. extrapolation.
factor de ajustamento da variância. variance adjustment factor.
fechado. closed.
filtragem. filtering.
filtragem linear. linear filtering.
filtro de Wiener. Wiener filter.
formalismo de máxima entropia. maximum entropy formalism.
fórmula de de Wijs. de Wijs formula.
fórmula de Krige. Krige's formula.
fórmula de Rodrigues. Rodrigues formula.
fractal. fractal.
frequência acumulada. cumulative frequency.
função. function.
função aleatória. random function.
função aleatória booleana. Boolean random function.
função aleatória conjuntamente ortogonal. jointly orthogonal random function.
função aleatória de diluição. dilution random function.
função aleatória de ruído branco. white random function.
função aleatória factorizável. factorable random function.
função aleatória intrínseca de ordem ν. intrinsic random function of order ν.
função aleatória ortogonal. orthogonal random function.
função aleatória semimarkoviana. semimarkovian random function.
função Beta. beta function.
função convexa. convex function.
função de amostragem. sample function.
função de autocorrelação. autocorrelation function.
função de base. basis function.
função de conjunto aditiva. additive set function.
função de distribuição. distribution function.
função de distribuição cumulativa. cumulative distribution function.
função de distribuição multidimensional. joint probability distribution.
função de Heaviside. Heaviside's function.
função delta de Dirac. Dirac delta function.
função densidade de probabilidade. probability density function.
função densidade de probabilidade bidimensional isofactorial. diagonal bivariate density function; isofactorial bivariate density function.
função densidade de probabilidade condicionada. conditional probability density function.

função densidade de probabilidade marginal. marginal probability density function.
função densidade de probabilidade multidimensional. joint probability density function.
função de penalização. loss function.
função de probabilidade. probability function.
função de recuperação. recovery function.
função de transferência. transfer function.
função de verosimilhança. likelihood function.
função em escada. step function.
função estritamente definida positiva. strictly positive definite function.
função estritamente definida positiva condicional. strictly conditionally positive definite function.
função estrutural. structural function; structure function.
função F. F-function.
função frequência. frequency function
função gama. gamma function.
função integrável. integrable function.
função integrável em relação a uma segunda função. integrable function with respect to a second function.
função intrínseca. intrinsic function.
função objectivo. objective function.
função ortogonal. orthogonal function.
função spline. spline function.
funções aleatórias conjuntamente factorizáveis. jointly factorable random function.
funções ortonormais. orthonormal function.
Geoestatística. geostatistics.
Geoestatística determinista. deterministic geostatistics.
Geoestatística linear. linear geostatistics.
Geoestatística não linear. nonlinear geostatistics.
Geoestatística não paramétrica. nonparametric geostatistics.
global. global.
grandeza regional. regional magnitude
grelha. grid.
harmónica. harmonic.
heteroscedástico. heteroscedastic.
heterotópico. heterotopic.
hipótese de permanência. conservation of law hypothesis.
hipótese intrínseca. intrinsic hypothesis.
histograma. histogram.
homogeneidade. homogeneity.
homoscedástico. homoscedastic.
impulso. impulse.
incerteza. uncertainty.
incrementos generalizados de ordem ν. generalized increments of order ν.
independência espacial. spatial independence.
independência estatística. statistical independence.
independência estocástica. stochastic independence.
independência linear. linear independence.
indicatriz. indicator.
infinitamente divisível. infinite divisibility.
infinitamente divisível para a união. infinite divisibility of a union.
ínfimo. infimum.
interior. interior.
interpolação. interpolation.
interpolador exacto. exact interpolator.
intervalo de confiança. confidence interval.
intervalo de constrangimentos. constraint interval.
intervalo de probabilidade. probability interval.
intervalo inter-quartílico. interquartile range; quartile deviation.
intrínseco. intrinsic.
isolinha. isoline; isopleth.
isotópica. isotopic.
isotropia. isotropy.
janela. window.
krigagem. kriging.
krigagem aleatória. random kriging.
krigagem da indicatriz. indicator kriging.
krigagem da indicatriz da mediana. median indicator kriging.
krigagem das componentes principais da indicatriz. indicator principal component kriging.
krigagem de blocos. block kriging.
krigagem de probabilidade. probability kriging.
krigagem disjuntiva. disjunctive kriging.
krigagem dos resíduos. residual kriging.
krigagem dual. dual kriging.
krigagem factorial. factorial kriging.
krigagem intrínseca. intrinsic kriging.
krigagem logarítmica. logarithmic kriging.
krigagem lognormal. lognormal kriging.
krigagem multigaussiana. multigaussian kriging.
krigagem múltipla da indicatriz. multiple indicator kriging.
krigagem ordinária. ordinary kriging.
krigagem pontual. point kriging; punctual kriging.
krigagem simples. simple kriging.
krigagem simples da indicatriz. single indicator kriging.
krigagem transitiva. transitive kriging.
krigagem universal. universal kriging.
lei dos grandes números. law of large numbers.
leptocúrtica. leptokurtic.
limite de confiança. confidence limit.
linha de uma matriz. row of a matrix.
linha de um determinante. row of a determinant.
local. local.
madograma. madogram.
malha. grid; lag.
matriz. matrix.
matriz de covariância. covariance matrix.
matriz diagonal. diagonal matrix.

matriz diagonalizada. diagonalized matrix.
matriz Hermitiana. Hermitian matrix.
matriz identidade. identity matrix.
matriz inversa. inverse of a matrix.
matriz quadrada. square matrix.
matriz simétrica. symmetric matrix.
matriz singular. singular matrix.
matriz transposta. transpose of a matrix.
média. average.
média aritmética. arithmetic average; arithmetic mean.
média geométrica. geometric average; geometric mean.
média harmónica. harmonic mean.
média móvel. moving linear average.
mediana. median.
medida. measure.
medida de Dirac. Dirac delta function.
medida de Hausdorff. Hausdorff measure.
medida de probabilidade. probability measure.
melhor estimador em erro quadrático médio. minimum mean square error estimator.
mesocúrtica. mesokurtic.
método da deriva externa. external drift method.
método das bandas rotativas. turning bands method.
método de Choleski. Choleski's method.
método de Monte Carlo. Monte Carlo method.
método dos círculos aleatórios. random tokens method; random coin method.
método dos mínimos quadrados. least squares method.
método dos multiplicadores de Lagrange. Lagrange's method of multipliers.
método transitivo. transitive method.
microergodicidade. microergodicity.
minimização da norma. minimum error norm estimator.
moda. mode.
modelo. model.
modelo bigaussiano. bigaussian model.
modelo de mosaico. mosaic model.
modelo de mudança de suporte. change of support model
modelo de variograma. semivariogram model.
modelo de variograma com efeito de cabeça de cavalo. hole effect semivariogram model.
modelo de variograma com expoente. power semivariogram model.
modelo gaussiano discretizado. discrete Gaussian model.
modelo hermitiano. Hermite model.
modelo linear de corregionalização. linear model of coregionalization.
modelo topoprobabilístico. topoprobabilistic model.
momento. moment.
momento de uma distribuição. moment of a distribution.
monoscópico. monoscopic.
movimento Browniano. Brownian motion process.
multidimensional. multivariate.
multiplicador de Lagrange. Lagrange multiplier.
não enviesado. unbiased.
não enviesamento condicional. conditional nonbias; conditional unbiasedness.
norma. norm.
número complexo conjugado. conjugate complex number.
núvem de correlação. cross-plot; scattergram; scatter-plot.
núvem de correlação a passo h. h-scattergram.
operador. operator.
ordem de uma matriz quadrada. order of a square matrix.
ordem de um determinante. order of a determinant.
panscópico. panscopic.
parâmetro convencional. conventional parameter.
partição de um conjunto. partition of a set.
passo. lag.
patamar. sill.
percentil. centile; percentile.
pivot. pivot.
platicúrtica. platykurtic.
polígono de influência. polygon of influence.
polígono de Thiessen. Thiessen polygon.
polígono de Voronoi. proximal polygon; Voronoi polygon.
polinómio de grau *n*. polynomial of degree *n*.
polinómio em várias variáveis. polynomial in several variables.
polinómio ortogonal. orthogonal polynomial.
polinómios de Hermite. Hermite polynomials.
polinómios de Laguerre. Laguerre polynomials.
poliscópico. polyscopic
ponto de acumulação. accumulation point.
população. population; universe.
precisão. precision.
previsão. prediction.
primeiro quartil. lower quartile.
princípio da característica regional. regional characteristic principle.
princípio das grandezas regionais. regional magnitude principle.
probabilidade. probability.
probabilidade condicionada. conditional probability.
probabilidade de transição. transition probability.
problema directo. direct problem; forward problem; normal problem.
problema inverso. inverse problem.
processo aleatório. random process.
processo aleatório conjuntamente ortogonal. jointly orthogonal random process.
processo aleatório factorizável. factorable random process.

processo aleatório ortogonal. orthogonal random process.
processo de Markov. Markov process.
processo de Poisson. Poisson process.
processo de Wiener. Wiener process.
processo de Wiener–Levy. Wiener–Levy process.
processo estocástico. stochastic process.
produto de matrizes. product of matrices.
produto de um escalar por uma matriz. product of a scalar and a matrix.
produto escalar. scalar product.
produto interno. inner product.
propriedade ergódica. ergodic property.
provas de Bernoulli. Bernoulli experiment.
quantil. fractile; quantile.
quartil. quartile.
quase-estacionário. quasi-stationary.
realização. realization.
recta de regressão. regression line.
recursividade. recursion.
regressão. regression.
regularização. regularization; smoothing.
relação de Cartier. Cartier's relation.
representação deslizante. sliding representation.
representação de uma função aleatória intrínsica. representation of an intrinsic random function.
reservas in situ. reserves in place.
reservas recuperáveis. recoverable reserves.
resíduo. residual.
resíduo optimal. optimal residual.
resíduo reduzido. reduced residual.
resistente. resistant.
resultado. outcome.
robustez. robust.
rodograma. rodogram.
ruído branco. white noise.
selectiva. selective.
semivariograma. demivariogram; semivariogram.
semivariograma cruzado. cross-semivariogram.
semivariograma (segundo Jowett & Matérn). serial variation function.
série temporal. time series.
significado objectivo. objective meaning.
simulação. simulation.
simulação condicional. conditional simulation.
simulação sequencial da indicatriz. sequential indicator simulation.
sistema completo de funções ortogonais. complete set of orthogonal functions.
sistema completo de polinómios ortogonais. complete set of orthogonal polynomials.
sistema determinado. regular system.
sistema não-singular. nonsingular system.
símbolo delta de Kronecker. Kronecker delta.
sobreestimação. overestimation.
subdivisão de Delaunay. Delaunay subdivision.
subespaço afim. linear manifold.
subestimação. underestimation.
sucessão de Cauchy. Cauchy sequence.
suporte. support.
suporte geométrico. geometric support.
tecelagem de Dirichlet. Dirichlet tesselation.
tecelagem de Voronoi. Voronoi tesselation.
técnica de jackknife. jackknife.
tendência. drift; trend.
teorema da aditividade. additivity theorem.
teorema de Bayes. Bayes' theorem.
teorema de Pitágoras. Pythagorean theorem.
teorema de projecção. projection theorem.
teorema do limite central. central limit theorem.
teorema ergódico. ergodic theorem.
teoria. theory.
terceiro quartil. upper quartile.
transformação. transformation.
transformação afim. affinity transformation.
transformação de segunda espécie. space transformation of the second kind.
transformação de primeira espécie. space transformation of the first kind.
transformação por ordenação crescente. rank order transform.
transformação uniforme. uniform transformation.
transformada. transform.
transformada de Fourier. Fourier transform.
transformada de Fourier inversa. Fourier synthesis.
transformada de Laplace. Laplace transform.
transformada rápida de Fourier. fast Fourier transform.
transitivo. transitive.
translacção de coeficientes autorizados. translation of permissible coefficients.
triangulação de Delaunay. Delaunay triangulation.
unidimensional. univariate.
universal. universal.
universo. universe.
validação. validation.
validação cruzada. cross-validation.
validação cruzada conjunta. joint cross-validation.
valor esperado. expected value.
valor estimado. estimate.
valor extremo aberrante. outlier.
valor médio. average value.
valor próprio. eigenvalue.
valor próprio de uma matriz. eigenvalue of a matrix.
variância. variance.
variância à priori. a priori variance.
variância de dispersão. dispersion variance.
variância de estimação. estimation variance.
variância de extensão. extension variance.
variância de krigagem. kriging variance.
variável aleatória. aleatory variable; random variable; variate.

variável aleatória contínua. continuous random variable.
variável aleatória de ordem 2. second order random variable.
variável aleatória discreta. discrete random variable.
variável estocástica. stochastic variable.
variável regionalizada. regionalized variable.
variedade. manifold.
variedade linear. linear manifold.
variografia. variography.
variograma. variogram.
variograma acumulado. graded semivariogram.
variograma cruzado. cross-variogram.
variograma cúbico. cubic semivariogram.
variograma da transformada por ordenação crescente. rank order variogram.
variograma de de Wijs. de Wijsian semivariogram.
variograma de ordem α. variogram of order α.
variograma embricado. nested semivariogram.
variograma esférico. spherical semivariogram.
variograma experimental. experimental semivariogram.
variograma experimental relativo. pairwise relative semivariogram.
variograma exponencial. exponential semivariogram.
variograma gaussiano. Gaussian semivariogram.
variograma generalizado. generalized semivariogram.
variograma global relativo. general relative semivariogram.
variograma linear. linear semivariogram.
variograma local relativo. local relative semivariogram.
variograma logarítmico. logarithmic semivariogram.
variograma médio. average semivariogram; mean semivariogram.
variograma parabólico. parabolic semivariogram.
variograma penta-esférico. penta-spherical semivariogram.
variograma pepítico. pure nugget effect semivariogram model.
variograma regional. regional semivariogram.
variograma regional probabilizado. probabilistic regional semivariogram.
variograma regularizado. regularized semivariogram.
variograma relativo. relative semivariogram.
variograma subjacente. underlying semivariogram.
variograma teórico. theoretical semivariogram.
vector. vector.
vectores ortonormados. orthonormal vector.
vector ortogonal. orthogonal vector.
vector próprio. eigenvector.
vizinhança. neighborhood.
vizinhança de krigagem. kriging neighborhood.
vizinhança deslizante. sliding neighborhood.
vizinhança de trabalho. working neighborhood.
zona de influência. zone of influence.

RUSSIAN to ENGLISH

σ–алгебра. σ–algebra.
абсолютная дисперсия. absolute dispersion.
автоковариация. autocovariance.
автокоррелограмма. autocorrelogram.
автокорреляционная функция. autocorrelation function.
автокорреляция. autocorrelation.
автокрайгинг. autokriging.
автоподобный. selfsimilar.
аддитивная теорема. additivity theorem.
аккумуляция. accumulation.
анализ Фурье. Fourier analysis.
анаморфоз. anamorphosis.
анаморфоз Гаусса. Gaussian anamorphosis.
анизотропия. anisotropy.
ансамбль. ensemble.
апостериорное распределение. posterior distribution.
аппроксимация кривой. curve fitting.
аппроксимация семивариограммы. semivariogram fitting.
априорная дисперсия. a priori variance.
априорное распределение. prior distribution.
асимметрия. skewness.
аффинная коррекция. affine correction.
аффинное преобразование. affinity transformation.
базисная функция. basis function.
балансовые запасы. recoverable reserves.
белый шум. white noise.
бесконечная делимость. infinite divisibility.
бесконечная делимость объединения. infinite divisibility of a union.
бета-распределение. beta distribution.
бета-функция. beta function.
бигауссовская модель. bigaussian model.
биекция. bijection.
бимодальное распределение. bimodal distribution.
биномиальное распределение. binomial distribution.
близкий полигон. proximal polygon.
блок-крайгинг. block kriging.
борелево поле. Borel field.
Броуновский процесс движения. Brownian motion process.
Булевская случайная функция. Boolean random function.
быстрое преобразование Фурье. fast Fourier transform.
вариограмма. variogram.
вариограмма порядка α. variogram of order α.
вариограмма рангого порядка. rank order variogram.
вариография. variography.
вектор. vector.
векторное подпространство. linear manifold.
векторное пространство. vector space.
вероятностная региональная семивариограмма. probabilistic regional semivariogram.
вероятность. probability.
вероятность перехода. transition probability.
верхний квартиль. upper quartile.
взаимная вариограмма. cross-variogram.
взаимная ковариация. cross-covariance.
взаимная ковариация индикатора. indicator cross-covariance.
взаимная корреляция. cross-correlation.
взаимная обоснованность. cross-validation.
взаимная семивариограмма. cross-semivariogram.
взаимное корреляционное поле. cross-plot.
видимая разрывность в начале координат. apparent discontinuity at the origin.
внутреннее произведение. inner product.
внутреннее свойство. intrinsic attribute.
внутренний. interior; intrinsic.
внутренний крайгинг. intrinsic kriging.
внутренний признак. intrinsic attribute.
внутренняя гипотеза. intrinsic hypothesis.
внутренняя дисперсия. intrinsic dispersion.
внутренняя корреляция. intrinsic correlation.
внутренняя случайная функция порядка ν. intrinsic random function of order ν.
внутренняя функция. intrinsic function.
временный ряд. time series.
выборка. sample.
выборочная функция. sample function.
выборочное дескриптивное пространство. sample description space.
выборочное пространство. sample space.
выбраковка. outlier.
выпуклая функция. convex function.
гамма-распределение. gamma distribution.
гамма-функция. gamma function.
гармоническая средняя. harmonic mean.
гармонический. harmonic.
г-диаграмма рассеивания. h-scattergram.
генеральная относительная семивариограмма. general relative semivariogram.
генеральный. universe.
геометрическая анизотропия. geometric anisotropy.
геометрическая ковариограмма. geometric covariogram.
геометрическая средняя. geometric mean.
геометрический объем. geometric support.
геометрическое поле. geometric field.
геометрическое среднее. geometric average.
геостатистика. geostatistics.
гетероскедастический. heteroscedastic.

гетеротопический. heterotopic.
Гильбертово пространство. Hilbert space.
гистограмма. histogram.
главная диагональ матрицы. main diagonal of a matrix; principal diagonal of a matrix.
главная компонента. principal component.
глобальная оценка. global estimation.
глобальные запасы. reserves in place.
глобальный. global.
гнездовая семивариограмма. nested semivariogram.
гнездовая структура. intermeshed structure; nested structure.
гомогенность. homogeneity.
гомологическая коразмерность. codimension.
гомоскедастический. homoscedastic.
градуирование. grading.
градуированная семивариограмма. graded semivariogram.
данная величина. datum.
данные. data.
двойственный крайгинг. dual kriging.
двухмерная переменная. bivariate.
двухмерное нормальное распределение. bivariate normal distribution.
двухмерное распределение. bivariate distribution.
двухпараметрическое логнормальное распределение. two-parameter lognormal distribution.
дегруппирование. declustering.
де-конволюция. deconvolution.
дельта-функция Дирака. Dirac delta function; impulse.
демивариограмма. demivariogram.
деструктуризация высокого содержания. high-grade destructuration.
детерминант. determinant.
детерминистская геостатистика. deterministic geostatistics.
дециль. decile.
диагонализируемая матрица. diagonalized matrix.
диагональная матрица. diagonal matrix.
диаграмма квантиль. q-q plot.
диаграмма рассеивания. scattergram
диаграмма розы. rose diagram.
диаграмма частоты. frequency diagram.
дизъюнктивный крайгинг. disjunctive kriging.
дискретная гауссовская модель. discrete Gaussian model.
дискретная случайная переменная. discrete random variable.
дисперсионный коэффициент выравнивания. variance adjustment factor.
дисперсия. variance.
дисперсия крайгинга. kriging variance.
дисперсия распространения. extension variance.
дисперсия среднего квадратического отклонения. dispersion variance.
доверительный интервал. confidence interval.
доверительный предел. confidence limit.
допустимые комбинации порядка ν. authorized combinations of order ν; permissible combinations of order ν.
допустимые коэффициенты порядка ν. permissible coefficients of order ν; authorized coefficients of order ν.
дрифт. drift.
дробная размерность. fractal dimension.
дырочный эффект. hole effect.
единичная матрица. identity matrix.
единичная функция Хевисайда. Heaviside's function.
единичный индикаторный крайгинг. single indicator kriging.
ёмкость Шоке. Choquet's capacity.
з-"score". z-score.
закон больших чисел. law of large numbers.
замкнутый. closed.
замощение Дирихле. Dirichlet tesselation.
замощение плоскости Вороного. Voronoi tesselation.
запаздывание. lag.
з-значение. z-score.
зона влияния. zone of influence.
зональная анизотропия. zonal anisotropy.
изолиния. isoline.
изоморфизм. bijection.
изоплета. isopleth.
изотопический. isotopic.
изотропия. isotropy.
индикатор. indicator.
индикатор ковариация. indicator covariance.
индикатор кокрайгинга. indicator cokriging.
индикатор крайгинг. indicator kriging.
интегральная область значений. integral range.
интегрируемая функция. integrable function.
интегрируемая функция по второй производной функции. integrable function with respect to a second function.
интервал вероятности. probability interval.
интервал влияния. range.
интервал условия. constraint interval.
интервальная функция распределения. cumulative distribution function.
интерквартильная широта. interquartile range.
интерполяция. interpolation.
инфимум. infimum.
информационный эффект. information effect.
имитация. simulation.
квадратная матрица. square matrix.
квазистационарный. quasi-stationary.
квантиль. fractile; quantile.
квартиль. quartile.
квартильное отклонение. quartile deviation.
клетка. Meijering's cell; Wigner–Seitz cell.
ковариация. covariance.
кокрайгинг. cokriging.
компактный. compact.
конвенциональный параметр. conventional parameter.
конволюция. convolution.

консистенция. consistency.
контур. contour.
конфигурация. configuration.
коррелограмма. correlogram.
коррелограмма взаимной корреляции. crosscorrelogram.
корреляция. correlation.
кофактор элемента в определителе. cofactor of an element in a determinant.
коэффициент автокорреляции. autocorrelation coefficient.
коэффициент асимметрии. coefficient of skewness.
коэффициент вариации. coefficient of variation.
коэффициент взаимной корреляции. cross-correlation coefficient.
коэффициент корреляции. correlation coefficient.
коэффициент корреляции Пирсона. Pearson correlation coefficient.
коэффициент корреляции Спирмана. Spearman correlation coefficient.
коэффициент эксцесса. coefficient of excess; coefficient of kurtosis.
крайгинг. kriging.
крайгинг вероятности. probability kriging.
крайгинг главных компонентов индикатора. indicator principal component kriging.
кратный индикаторный крайгинг. multiple indicator kriging.
кривая селективности. selectivity curve.
кривая частоты. frequency curve.
критерии допустимости. permissibility criteria.
критерий еффективности. objectivity criterion.
кубическая семивариограмма. cubic semivariogram.
линейная геостатистика. linear geostatistics.
линейная зависимость. linear dependence.
линейная комбинация переменных. linear combination of variables.
линейная независимость. linear independence.
линейная семивариограмма. linear semivariogram.
линейная фильтрация. linear filtering.
линейное многообразие. linear manifold.
линейное пространство. linear space.
линия регрессии. regression line.
логарифмическая семивариограмма. logarithmic semivariogram.
логарифмический крайгинг. logarithmic kriging.
логнормальная коррекция. lognormal correction.
логнормальное распределение. lognormal distribution.
логнормальный крайгинг. lognormal kriging.
локализованная переменная. regionalized variable.
локальная оценка. local estimation.
локальная относительная семивариограмма. local relative semivariogram.
локальный. local.
мадограмма. madogram.
маргинальное распределение. marginal distribution.
марковский процесс. Markov process.
математическое ожидание. expected value.
матрица. matrix.
матрица ковариации. covariance matrix.
матрица Эрмита. Hermitian matrix.
медиана. median.
медианный индикаторный крайгинг. median indicator kriging.
мезовершинная. mesokurtic.
мера. measure.
мера вероятности. probability measure.
мера Гаусдорфа. Hausdorff measure.
метка размытого множества. fuzzy indicator data.
метод внешнего дрейфа. external drift method.
метод вращающихся полосок. turning bands method.
метод Монте-Карло. Monte Carlo method.
метод наименьших квадратов. least squares method.
метод неопределенных множителей Лагранжа. Lagrange's method of multipliers.
метод случайного выигрыша поверхности. random coin method; random tokens method.
метод статистических испытаний. Monte Carlo method.
метод Холеского. Choleski's method.
метрическое пространство. metric space.
микроэргодичность. microergodicity.
многогауссовский крайгинг. multigaussian kriging.
многомерная случайная переменная. multivariate.
многомерное нормальное распределение. multinormal distribution.
многомерное распределение. multivariate distribution.
многомодальное распределение. multimodal distribution.
многочлен от многих переменных. polynomial in several variables.
многочлен степени н. polynomial of degree *n*.
многочлены Чебышева-Лагерра. Laguerre polynomials.
многочлены Эрмита. Hermite polynomials.
множество дробной размерности. fractal.
множитель Лагранжа. Lagrange multiplier.
мода. mode.
моделирование. simulation.
моделирование последовательного индикатора. sequential indicator simulation.
модель. model.
модель дырочного эффекта семивариограммы. hole effect semivariogram model.
модель изменения суппорта. change of support model.
модель семивариограммы. semivariogram model.
модель семивариограммы чистого эффекта самородков. pure nugget effect semivariogram model.
модель Эрмита. Hermite model.
мозаичная модель. mosaic model.
момент. moment.
момент распределения. moment of a distribution.
моноскопический. monoscopic.
накопленная частота. cumulative frequency.

накопление. accumulation.
направление. trend.
недооценивание. underestimation.
недооценка. underestimation.
нелинейная геостатистика. nonlinear geostatistics.
неопределенность. uncertainty.
непараметрическая геостатистика. nonparametric geostatistics.
непрерывная случайная величина. continuous random variable.
непрямая логнормальная коррекция. indirect lognormal correction.
неравенство Бесселя. Bessel inequality.
неравенство Буняковского. Buniakovski inequality.
неравенство Коши-Шварца. Cauchy–Schwarz inequality.
неравенство Шварца. Schwarz inequality.
несмещенный. unbiased.
неточные данные. soft data.
нецентрированная ковариация. noncentered covariance.
неэргодическая экспериментальная ковариация. nonergodic experimental covariance.
нижнее-верхнее (НВ) треугольное разложение. lower-upper (LU) decomposition.
нижний квартиль. lower quartile.
норма. norm.
нормальное гауссовское распределение. Gaussian distribution.
нормальное преобразование ядра оператора. normal score transform.
нормальное распределение. normal distribution.
нормальные уравнения. normal equations.
носитель локализованной переменной. support.
область. domain.
обобщенная ковариация порядка v. generalized covariance of order v.
обобщенная семивариограмма. generalized semivariogram.
обобщенные приросты порядка v. generalized increments of order v.
обоснованность. validation.
обратная задача. inverse problem.
обращение матрицы. inverse of a matrix.
объем выборки. sample size.
обыкновенный крайгинг. ordinary kriging.
ограничение. constraint effect.
одномерная переменная. univariate.
однородность. homogeneity.
окно. window.
окрестность. neighborhood.
окрестность крайгинга. kriging neighborhood.
оператор. operator.
операционное понятие. operational concept.
определитель. determinant.
определитель матрицы. determinant of a matrix.
оптимальный остаток. optimal residual.
ортогональная случайная функция. orthogonal random function.
ортогональная функция. orthogonal function.
ортогональный вектор. orthogonal vector.
ортогональный многочлен. orthogonal polynomial.
ортогональный случайный процесс. orthogonal random process.
ортонормальная функция. orthonormal function.
ортонормальный вектор. orthonormal vector.
особенная матрица. singular matrix.
остаток. residual.
остаточный крайгинг. residual kriging.
островершинное. leptokurtic.
открытый. open.
относительная семивариограмма. relative semivariogram.
относительный эффект самородков. relative nugget effect.
отрицательное биномиальное распределение. negative binomial distribution.
оценивание. estimation.
оценка. estimation.
оценка "jackknife". jackknife.
оценка дисперсии. estimation variance.
оценка максимального правдоподобия. maximum likelihood estimator.
оценка нормы минимальной ошибки. minimum error norm estimator.
оценка ошибки минимального квадрата средней величины. minimum mean square error estimator.
оценка складного ножа. jackknife.
оценка типа Э. E-type estimate.
ошибка. error.
панскопический. panscopic.
параболическая семивариограмма. parabolic semivariogram.
пентасферическая семивариограмма. penta-spherical semivariogram.
передаточная функция. transfer function.
переменная. aleatory variable.
переоценивание. overestimation.
переоценка. overestimation.
плосковершинное. platykurtic.
подразделение Делона. Delaunay subdivision.
поле. field.
поле рассеивания. scatterplot.
полигон влияния. polygon of influence.
полигон Вороного. Voronoi polygon.
полигон Тиссена. Thiessen polygon.
полископический. polyscopic.
полное множество ортогональных многочленов. complete set of orthogonal polynomials.
полное множество ортогональных функций. complete set of orthogonal functions.
полное пространство. complete space.
положительный определенный. positive definite.
полумарковская случайная функция. semimarkovian

random function.
попарно относительная семивариограмма. pairwise relative semivariogram.
популяция. population.
порог. sill.
порядок квадратной матрицы. order of a square matrix.
порядок определителя. order of a determinant.
последовательность Коши. Cauchy sequence.
преапостериорное распределение. pre-posterior distribution.
предельная точка. accumulation point.
представление внутренней случайной функции. representation of an intrinsic random function.
преобразование. transform; transformation.
преобразование Лапласа. Laplace transform.
преобразование пространства второй степени. space transformation of the second kind.
преобразование пространства первой степени. space transformation of the first kind.
преобразование рангого порядка. rank order transform.
преобразование Фурье. Fourier transform.
приведенный остаток. reduced residual.
принцип региональной величины. regional magnitude principle.
принцип региональной характеристики. regional characteristic principle.
прогноз. prediction.
произведение матриц. product of matrices.
произведение матрицы на число. product of a scalar and a matrix.
пропорциональный эффект. proportional effect.
простой крайгинг. simple kriging.
пространственная зависимость. spatial dependence.
пространственная независимость. spatial independence.
пространство вероятности. probability space.
пространство выборок. event space; sample space.
пространство с операцией внутреннего произведения. inner product space.
процентиль. centile; percentile.
процесс Винера. Wiener process.
процесс Винера-Леви. Wiener–Levy process.
процесс по параболе. parabolic behavior.
прямая задача. direct problem; normal problem.
Пуассоновский процесс. Poisson process.
рабочая окрестность. working neighborhood.
рабочий масштаб. working scale.
равенство Парсеваля. Parseval's equation.
равномерное кондиционирование. uniform conditioning.
равномерное преобразование. uniform transformation.
равномерное распределение. uniform distribution.
разбиение множества. partition of a set.
разделение Вороного. Voronoi division.
размерность Гаусдорфа-Безиковича. Hausdorff–Besicovitch dimension.
размерность подобия. similarity dimension.
размер подобия. similarity dimension.
разнообразие. manifold.
разрешающий. pivot.
распределение. distribution.
распределение вероятности. probability distribution.
распределение Гаусса. Gaussian distribution.
распределение Пуассона. Poisson distribution.
распределение совместной вероятности. joint probability distribution.
расслоенный случайный выбор. stratified random sampling.
расстояние между квартилями. interquartile range.
рациональное выражение. rational expression.
реализация. outcome; realization.
региональная величина. regional magnitude
региональная семивариограмма. regional semivariogram.
региональная характеристика. regional characteristic.
регрессия. regression.
регулярная система. nonsingular system; regular system.
регуляризация. regularization.
регуляризованная семивариограмма. regularized semivariogram.
рекурсия. recursion.
робастный. robust.
родограмма. rodogram.
свойство эргодичности. ergodic property.
сглаживание. smoothing.
селективный. selective.
семивариограмма. semivariogram.
семивариограмма Гаусса. Gaussian semivariogram.
семивариограмма де Вийса. de Wijsian semivariogram.
семивариограмма для остатков локализованных переменных. underlying semivariogram.
сеть. grid.
сигма-поле. sigma-field.
силь. sill.
сильная стационарность. strong stationarity.
символ Кронекера. Kronecker delta.
симметрическая матрица. symmetric matrix.
синтез Фурье. Fourier synthesis.
синхронная задача. forward problem.
скалярное произведение. scalar product.
скользящая окрестность. sliding neighborhood.
скользящее линейное среднее. moving linear average.
скользящее представление. sliding representation.
слабая стационарность. weak stationarity.
случайная величина. aleatory variable; variate.
случайная переменная. random variable.
случайная переменная второго порядка. second order random variable.
случайная функция. random function.
случайная функция белого процесса. white random function.
случайная функция разбавления. dilution random function.

случайное множество. random set.
случайное поле. random field.
случайное поле второго порядка. second order random field.
случайный крайгинг. random kriging.
случайный процесс. random process.
случайный эксперимент. trial; random experiment.
смещение. bias.
собственное значение. eigenvalue.
собственное значение матрицы. eigenvalue of a matrix.
собственный вектор. eigenvector.
событие. event.
совместная взаимная обоснованность. joint cross-validation.
совместное локализирование. coregionalization.
совместное моделирование. coconditional simulation; conditional cosimulation; cosimulation.
совместно ортогональная случайная функция. jointly orthogonal random function.
совместно ортогональный случайный процесс. jointly orthogonal random process.
совместно факторизуемая случайная функция. jointly factorable random function.
совокупность. population.
соответствующая линейная модель соответствующего локализирования. linear model of coregionalization.
соответствующее моделирование. coconditional simulation; conditional; cosimulation; cosimulation.
соответствующий крайгинг. cokriging.
соотношение Картье. Cartier's relation.
сопротивляющийся. resistant.
сопряженное комплексное число. conjugate complex number.
сохранение гипотезы закона. conservation of law hypothesis.
сплайн-функция. spline function.
способность к автокрайгингу. autokrigeability.
среднее. average.
среднее арифметическое. arithmetic average.
среднее гармоническое. harmonic mean.
среднее геометрическое. geometric mean.
среднее значение. average value.
среднеквадратичная ошибка. mean square error.
средняя арифметическая величина. arithmetic mean.
средняя величина. average value; mean.
средняя семивариограмма. average semivariogram; mean semivariogram.
стандартная ошибка. standard error.
стандартное з-значение. standard score.
стандартное нормальное распределение. standard normal distribution.
стандартное отклонение. standard deviation.
статистика. estimator; statistic.
статистика максимального правдоподобия. maximum likelihood statistic.
статистическая зависимость. statistical dependence.
статистическая независимость. statistical independence.
статистическая оценка. estimate.
стационарность. stationarity.
стационарный. stationary.
стационарный в широком смысле. stationary in the wide sense.
стационарный конечного порядка. stationary of finite order.
степенная модель семивариограммы. power semivariogram model.
столбец данных индикатора. indicator data column.
столбец детерминанта. column of a determinant.
столбец матрицы. column of a matrix.
стохастическая зависимость. stochastic dependence.
стохастическая независимость. stochastic independence.
стохастическая переменная. stochastic variable.
стохастический процесс. stochastic process.
стратифицированная анизотропия. stratified anisotropy.
строго положительная определенная функция. strictly positive definite function.
строго условно положительная определенная функция. strictly conditionally positive definite function.
строка матрицы. row of a matrix.
строка определителя. row of a determinant.
структура. structure.
структуризация высокого содержания. high-grade structuration.
структурная функция. structural function.
структурный анализ. structural analysis.
ступенчатая функция. step function.
суппорт. support.
сферическая семивариограмма. spherical semivariogram.
теорема Байеса. Bayes' theorem.
теорема Пифагора. Pythagorean theorem.
теорема проекции. projection theorem.
теоретическая семивариограмма. theoretical semivariogram.
теория. theory.
топовероятностная модель. topoprobabilistic model.
точечный крайгинг. point kriging; punctual kriging.
точка накопления. accumulation point.
точная стационарность. strict stationarity.
точность. accuracy; precision.
точные данные. hard data.
точный интерполятор. exact interpolator.
транзитивная ковариограмма. transitive covariogram.
транзитивный. transitive.
транзитивный крайгинг. transitive kriging.
транзитивный метод. transitive method.
трансляция допустимых коэффициентов. translation of permissible coefficients.
транспонировать матрицу. transpose of a matrix.
т-распределение Сихеля. Sichel t distribution.
тренд. trend.
трехпараметрическое логнормальное распределение. three-parameter lognormal distribution.

триангуляция Делона. Delaunay triangulation.
универсальное условие. universal condition.
универсальный. universal.
универсальный крайгинг. universal kriging.
унимодальное распределение. unimodal distribution.
условная вероятность. conditional probability.
условная несмещенность. conditional unbiasedness.
условное моделирование. conditional simulation.
условное не-смещение. conditional nonbias.
условное ожидание. conditional expectation.
условное распределение. conditional distribution.
условное смещение. conditional bias.
условно положительная определенная порядка ν. conditionally positive definite of order ν.
устойчивость объединения. union stability.
устойчивый. robust.
факториальный крайгинг. factorial kriging.
факторизуемая случайная функция. factorable random function.
факторизуемый случайный процесс. factorable random process.
факторный анализ. factor analysis.
фильтрация. filtering.
фильтр Винера. Wiener filter.
фильтрирование. filtering.
формализм максимальной энтропии. maximum entropy formalism.
формула де Вийса. de Wijs formula.
формула Криге. Krige's formula.
формула Родригесса. Rodrigues formula.
фрактал. fractal.
функция. function.
функция аддитивного множества. additive set function.
функция вероятности. probability function.
функция выборки. sample function.
функция извлечения. recovery function.
функция плотности вероятности. probability density function.
функция плотности диагональной двумерной переменной. diagonal bivariate density function.
функция плотности изофакториальной двумерной случайной переменной. isofactorial bivariate density function.
функция плотности маргинальной вероятности. marginal probability density function.
функция плотности совместной вероятности. joint probability density function.
функция плотности условной вероятности. conditional probability density function.
функция-потерь. loss function.
функция правдоподобия. likelihood function.
функция распределения. distribution function.
функция сериалной дисперсии. serial variation function.
функция структуры. structure function
функция ф. F-function.
функция Хевисайда. Heaviside's function.
функция частоты. frequency function.
ф-функция. F-function.
характеристическое уравнение матрицы. characteristic equation of a matrix.
целевая функция. objective function.
целевое значение. objective meaning.
целевой смысл. objective meaning.
центральная граничная теорема. central limit theorem.
центрированная ковариация. centered covariance.
часть поля. window.
чистый эффект самородков. pure nugget effect.
эвклидово пространство. euclidean space.
эксперимент. experiment.
экспериментальная ковариация. experimental covariance.
экспериментальная семивариограмма. experimental semivariogram.
эксперимент Берноулли. Bernoulli experiment.
экспоненциальная семивариограмма. exponential semivariogram.
экспоненциальное распределение. exponential distribution.
экстраполирование. extrapolation.
экстраполяция. extrapolation.
эксцесс. kurtosis.
элемент детерминанта. element of a determinant.
элемент матрицы. element of a matrix.
эллипс Криге. Krige's ellipse.
эллиптическая анизотропия. elliptical anisotropy.
энтропия. entropy.
эргодическая теорема. ergodic theorem.
эргодичество. ergodicity.
эффект носителя множества. support effect.
эффект отсеивания. screen effect.
эффект просеивания. screen effect.
эффект самородков. nugget effect.
ячейка Вигнера-Зайтца. Wigner–Seitz cell.
ячейка Мейеринга. Meijering's cell.

SPANISH to ENGLISH

abierto. open.
acumulación. acumulation; grading.
achatado. platykurtic.
ajuste de una curva. curve fitting.
ajuste de un semivariograma. semivariogram fitting.
alcance. range (b).
alcance integral. integral range.
álgebra σ. σ–algebra.
amplitud. range (a).
amplitud intercuartílica. interquartile range.
análisis de Fourier. Fourier analysis.
análisis estructural. structural analysis.
análisis factorial. factor analysis.
anamorfosis. anamorphosis.
anamorfosis gaussiana. Gaussian anamorphosis.
anisotropía. anisotropy.
anisotropía elíptica. elliptical anisotropy.
anisotropía estratificada. stratified anisotropy.
anisotropía geométrica. geometric anisotropy.
anisotropía zonal. zonal anisotropy.
armónica. harmonic.
asimetría. skewness.
atributo intrínseco. intrinsic attribute.
autocorrelación. autocorrelation.
autocorrelograma. autocorrelogram.
autocovarianza. autocovariance.
autohomotético. selfsimilar.
autokrigeabilidad. autokrigeability.
autokrigeage. autokriging.
autosemejante. selfsimilar.
autovalor. eigenvalue.
autovalor de una matriz. eigenvalue of a matrix.
autovector. autovector.
bivariable. bivariate.
biyección. bijection.
campo. field.
campo aleatorio. random field.
campo aleatorio de segundo orden. second order random field.
campo de Borel. Borel field.
campo geométrico. geometric field.
campo sigma. sigma-field.
capacitancia de Choquet. Choquet's capacity.
característica regional. regional characteristic; regional magnitude.
celda de Meijering. Meijering's cell.
celda de Wigner-Seitz. Wigner–Seitz cell.
cerrado. closed.
codimensión. codimension.
coeficiente de asimetría. coefficient of skewness.
coeficiente de autocorrelación. autocorrelation coefficient.
coeficiente de correlación. correlation coefficient.
coeficiente de correlación cruzada. cross-correlation coefficient.
coeficiente de correlación de Pearson. Pearson correlation coefficient.
coeficiente de correlación de Spearman. Spearman correlation coefficient.
coeficiente de curtosis. coefficient of kurtosis.
coeficiente de exceso. coefficient of excess.
coeficiente de variación. coefficient of variation.
coeficientes autorizados de orden ν. authorized coefficients of order ν; permissible coefficients of order ν.
cofactor de un elemento de un determinante. cofactor of an element in a determinant.
cokrigeado. cokriging.
cokrigeage. cokriging.
cokrigeage de indicadores. indicator cokriging.
columna de datos de un indicador. indicator data column.
columna de una matriz. column of a matrix.
columna de un determinante. column of a determinant.
combinaciones autorizadas de orden ν. authorized combinations of order ν; permissible combinations of order ν.
combinación lineal de variables. linear combination of variables.
compacto. compact.
componente principal. principal component.
comportamiento parabólico. parabolic behavior.
concepto operativo. operational concept.
condicionalmente definido positivo de orden v. conditionally positive definite of order v.
condicionamiento uniforme. uniform conditioning.
configuración. configuration.
conjunto. ensemble.
conjunto aleatorio. random set.
conjunto completo de funciones ortogonales. complete set of orthogonal functions.
conjunto completo de polinomios ortogonales. complete set of orthogonal polynomials.
consistencia. consistency.
convolución. convolution.
corrección afín. affine correction.
corrección lognormal. lognormal correction.
corrección lognormal indirecta. indirect lognormal correction.
corregionalización. coregionalization.

correlación. correlation.
correlación cruzada. cross-correlation.
correlación intrínseca. intrinsic correlation.
correlograma. correlogram.
correlograma cruzado. crosscorrelogram.
cosimulación. cosimulation.
cosimulación condicionada. coconditional simulation; conditional cosimulation.
covarianza. covariance.
covarianza centrada. centered covariance.
covarianza cruzada. cross-covariance.
covarianza cruzada de indicadores. indicator cross-covariance.
covarianza de indicadores. indicator covariance.
covarianza experimental. experimental covariance.
covarianza experimental no ergódica. nonergodic experimental covariance.
covarianza generalizada de orden ν. generalized covariance of order ν.
covarianza no centrada. noncentered covariance.
covariograma geométrico. geometric covariogram.
covariograma transitivo. transitive covariogram.
criterio de objetividad. objectivity criterion.
criterios de permisibilidad. permissibility criteria.
cuadrícula. grid.
cuantil. quantile.
cuartil. quartile.
cuartil inferior. lower quartile.
cuartil superior. upper quartile.
cuasi-estacionario. quasi-stationary.
curtosis. kurtosis.
curva de frecuencias. frequency curve.
curva de nivel. contour.
curva de selectividad. selectivity curve.
dato. datum (b).
datos. data.
datos de un indicador vago. fuzzy indicator data.
datos fiables. hard data.
datum. datum (a).
decil. decile.
deconvolución. deconvolution.
definida positiva. positive definite.
delta de Kronecker. Kronecker delta.
dependencia espacial. spatial dependence.
dependencia estadística. statistical dependence.
dependencia estocástica. stochastic dependence.
dependencia lineal. linear dependence.
desaglomeración. declustering.
desestructuración de las leyes altas. high grade destructuration.
desfasamiento. lag.
desigualdad de Bessel. Bessel inequality.
desigualdad de Buniakovski. Buniakovski inequality.
desigualdad de Cauchy-Schwarz. Cauchy–Schwarz inequality.
desigualdad de Schwarz. Schwarz inequality.
desviación cuartílica. quartile deviation.
desviación estándard. standard deviation.
determinante. determinant.
determinante de una matriz. determinant of a matrix.
diagonal principal de una matriz. main diagonal of a matrix; principal diagonal of a matrix.
diagrama de cuantiles. q-q plot.
diagrama de dispersión. cross-plot; scattergram; scatterplot.
diagrama de dispersión tipo h. h-scattergram.
diagrama de frecuencias. frequency diagram.
diagrama de roseta. rose diagram.
dimensión de Hausdorff-Besicovitch. Hausdorff–Besicovitch dimension.
dimensión de similitud. similarity dimension.
dimensión fractal. fractal dimension.
discontinuidad aparente en el origen. apparent discontinuity at the origin.
dispersión absoluta. absolute dispersion.
dispersión intrínseca. intrinsic dispersion.
distribución. distribution.
distribución a posteriori. posterior distribution.
distribución a priori. prior distribution.
distribución beta. beta distribution.
distribución bimodal. bimodal distribution.
distribución binomial. binomial distribution.
distribución binomial negativa. negative binomial distribution.
distribución bivariable. bivariate distribution.
distribución condicionada. conditional distribution.
distribución de Gauss. Gaussian distribution.
distribución de Poisson. Poisson distribution.
distribución de probabilidad. probability distribution.
distribución exponencial. exponential distribution.
distribución gamma. gamma distribution.
distribución lognormal. lognormal distribution.
distribución lognormal de dos parámetros. two-parameter lognormal distribution.
distribución lognormal de tres parámetros. three-parameter lognormal distribution.
distribución marginal. marginal distribution.
distribución multimodal. multimodal distribution.
distribución multinormal. multinormal distribution.
distribución multivariable. multivariate distribution.
distribución normal. normal distribution.
distribución normal bivariable. bivariate normal distribution.
distribución normal estandardizada. standard normal distribution.
distribución normal estandarizada. standard normal distribution.
distribución pre-posterior. pre-posterior distribution.
distribución t de Sichel. Sichel t distribution.
distribución uniforme. uniform distribution.
distribución unimodal. unimodal distribution.
divisibilidad infinita. infinite divisibility.

divisibilidad infinita de una unión. infinite divisibility of a union.
división de Voronoi. Voronoi division.
dominio. domain.
ecuación característica de una matriz. characteristic equation of a matrix.
ecuación de Parseval. Parseval's equation.
ecuaciones normales. normal equations.
efecto agujero. hole effect.
efecto de pepita puro. pure nugget effect.
efecto de pepita relativo. relative nugget effect.
efecto de proporcionalidad. proportional effect.
efecto información. information effect.
efecto pantalla. screen effect.
efecto pepita. nugget effect.
efecto restricción. constraint effect.
efecto soporte. support effect.
elemento de una matriz. element of a matrix.
elemento de un determinante. element of a determinant.
elipse de Krige. Krige's ellipse.
ensayo. trial.
entorno. neighborhood.
entorno de krigeage. kriging neighborhood.
entorno de trabajo. working neighborhood.
entorno móvil. sliding neighborhood.
entropía. entropy.
ergodicidad. ergodicity.
error. error.
error cuadrático medio. mean square error.
error estándard. standard error.
escala de trabajo. working scale.
espacio completo. complete space.
espacio de Hilbert. Hilbert space.
espacio euclidiano. Euclidean space.
espacio lineal. linear space.
espacio métrico. metric space.
espacio muestral. event space; sample description space; sample space.
espacio probabilístico. probability space.
espacio producto interior. inner product space.
espacio vectorial. vector space.
estabilidad de una unión. union stability.
estacionariedad. stationarity.
estacionariedad débil. weak stationarity.
estacionariedad estricta. strict stationarity; strong stationarity.
estacionario. stationary.
estacionario de orden finito. stationary of finite order.
estacionario en sentido amplio. stationary in the wide sense.
estadístico. statistic.
estadístico de máxima verosimilitud. maximum likelihood statistic.
estimación. estimation.
estimación global. global estimation.
estimación local. local estimation.
estimador. estimator.
estimador de máxima verosimilitud. maximum likelihood estimator.
estimador de mínima norma del error. minimum error norm estimator.
estimador de mínimo error cuadrático medio. minimum mean square error estimator.
estimador de tipo E. E-type estimate.
estructura. structure.
estructuración de las leyes altas. high-grade structuration.
estructura compuesta. intermeshed structure; nested structure.
estructura imbricada. intermeshed structure; nested structure.
evento. event.
exactitud. accuracy.
experimento. experiment.
experimento aleatorio. random experiment.
experimento de Bernoulli. Bernoulli experiment.
expresión racional. rational expression.
extrapolación. extrapolation.
factor de ajuste de la varianza. variance adjustment factor.
factorización LU. lower-upper (LU) decomposition.
fila de una matriz. row of a matrix.
fila de un determinante. row of a determinant.
filtrado. filtering.
filtro de Wiener. Wiener filter.
filtro lineal. linear filtering.
formalismo de máxima entropía. maximum entropy formalism.
fórmula de de Wijs. de Wijs formula.
fórmula de Krige. Krige's formula.
fórmula de Rodrigues. Rodrigues formula.
fractal. fractal.
fractil. fractile.
frecuencia acumulada. cumulative frequency.
función. function.
función aleatoria. random function.
función aleatoria de Bool. Boolean random function.
función aleatoria de dilución. dilution random function.
función aleatoria de ruido blanco. white random function.
función aleatoria factorizable. factorable random function.
función aleatoria intrínseca de orden ν. intrinsic random function of order ν.
función aleatoria ortogonal. orthogonal random function.
función aleatoria semimarkoviana. semimarkovian random function.
función aleatoria simultáneamente factorizable. jointly factorable random function.

función aleatoria simultáneamente ortogonal. jointly orthogonal random function.
función beta. beta function.
función convexa. convex function.
función de autocorrelación. autocorrelation function.
función de base. basis function.
función de conjunto aditiva. additive set function.
función de densidad de probabilidad. probability density function.
función de densidad de probabilidad condicionada. conditional probability density function.
función de densidad de probabilidad conjunta. joint probability density function.
función de densidad diagonal bivariable. diagonal bivariate density function.
función de densidad isofactorial bivariable. isofactorial bivariate density function.
función de densidad marginal de probabilidad. marginal probability density function.
función de distribución. distribution function.
función de distribución acumulada. cumulative distribution function.
función de distribución de probabilidad conjunta. joint probability distribution.
función de estructura. structure function.
función de frecuencia. frequency function.
función de Heaviside. Heaviside's function.
función delta de Dirac. Dirac delta function.
función de muestreo. sample function.
función de pérdida. loss function.
función de probabilidad. probability function.
función de recuperación. recovery function.
función de transferencia. transfer function.
función de variación en serie. serial variation function.
función de verosimilitud. likelihood function.
función escalón. step function.
función estrictamente definida condicionalmente positiva. strictly conditionally positive definite function.
función estrictamente definida positiva. strictly positive definite function.
función estructural. structural function.
función F. F-function.
función gamma. gamma function.
función integrable. integrable function.
función integrable con respecto a una segunda función. integrable function with respect to a second function.
función intrínseca. intrinsic function.
función objetivo. objective function.
función ortogonal. orthogonal function.
función ortonormal. orthonormal function.
geoestadística. geostatistics.
geoestadística determinística. deterministic geostatistics.
geoestadística lineal. linear geostatistics.
geoestadística no lineal. nonlinear geostatistics.
geoestadística no paramétrica. nonparametric geostatistics.
global. global.
grilla. grid.
heteroscedástico. heteroscedastic.
heterotópico. heterotopic.
hipótesis de conservación de la ley. conservation of law hypothesis.
hipótesis intrínseca. intrinsic hypothesis.
histograma. histogram.
homogeneidad. homogeneity.
homoscedástico. homoscedastic.
impulso. impulse.
incertidumbre. uncertainty.
incrementos generalizados de orden ν. generalized increments of order ν.
independencia espacial. spatial independence.
independencia estadística. statistical independence.
independencia estocástica. stochastic independence.
independencia lineal. linear independence.
indicador. indicator.
ínfimo. infimum.
información inexacta. soft data.
insesgado. unbiased.
intento. trial.
interior. interior.
interpolación. interpolation.
interpolador exacto. exact interpolator.
intervalo de confianza. confidence interval.
intervalo de la restricción. constraint interval.
intervalo de probabilidad. probability interval.
intrínseco. intrinsic.
isolínea. isoline.
isopleta. isopleth.
isotópico. isotopic.
isotropía. isotropy.
jackknife. jackknife.
krigeado. kriging.
krigeage. kriging.
krigeage aleatorio. random kriging.
krigeage de la mediana de un indicador. median indicator kriging.
krigeage de las componentes principales de los indicadores. indicator principal component kriging.
krigeage de probabilidad. probability kriging.
krigeage de un indicador. indicator kriging.
krigeage de un solo indicador. single indicator kriging.
krigeage de varios indicadores. multiple indicator kriging.
krigeage disyuntivo. disjunctive kriging.
krigeage dual. dual kriging.
krigeage factorial. factorial kriging.
krigeage intrínseco. intrinsic kriging.
krigeage logarítmico. logarithmic kriging.
krigeage lognormal. lognormal kriging.

krigeage multigaussiano. multigaussian kriging.
krigeage ordinario. ordinary kriging.
krigeage por bloques. block kriging.
krigeage puntual. point kriging; punctual kriging.
krigeage residual. residual kriging.
krigeage simple. simple kriging.
krigeage transitivo. transitive kriging.
krigeage universal. universal kriging.
leptocúrtico. leptokurtic.
ley de los grandes números. law of large numbers.
límite de confianza. confidence limit.
local. local.
madograma. madogram.
malla. grid.
matriz. matrix.
matriz cuadrada. square matrix.
matriz de covarianza. covariance matrix.
matriz de Hermite. Hermitian matrix.
matriz diagonal. diagonal matrix.
matriz diagonalizada. diagonalized matrix.
matriz identidad. identity matrix.
matriz inversa. inverse of a matrix.
matriz simétrica. symmetric matrix.
matriz singular. singular matrix.
matriz transpuesta. transpose of a matrix.
media. mean.
media aritmética. arithmetic mean.
media armónica. harmonic mean.
media geométrica. geometric mean.
media lineal móvil. moving linear average.
mediana. median.
medida. measure.
medida de Hausdorff. Hausdorff measure.
medida de probabilidad. probability measure.
meseta. sill.
mesocúrtico. mesokurtic.
método de Choleski. Choleski's method.
método de las bandas rotantes. turning bands method.
método de las monedas aleatorias. random coin method; random tokens method.
método de la tendencia externa. external drift method.
método de los multiplicadores de Lagrange. Lagrange's method of multipliers.
método de mínimos cuadrados. least squares method.
método de Monte Carlo. Monte Carlo method.
método transitivo. transitive method.
microergodicidad. microergodicity.
moda. mode.
modelo. model.
modelo bigaussiano. bigaussian model.
modelo de cambio de soporte. change of support model.
modelo de Hermite. Hermite model.
modelo de mosaico. mosaic model.
modelo de semivariograma potencial. power semivariogram model.
modelo de un semivariograma. semivariogram model.
modelo gaussiano discreto. discrete Gaussian model.
modelo lineal de corregionalización. linear model of coregionalization.
modelo topoprobabilístico. topoprobabilistic model.
momento. moment.
momento de una distribución. moment of a distribution.
monoscópico. monoscopic.
movimiento browniano. Brownian motion process.
muestra. sample.
muestreo aleatorio estratificado. stratified random sampling.
multiplicador de Lagrange. Lagrange multiplier.
multivariable. multivariate.
norma. norm.
número complejo conjugado. conjugate complex number.
operador. operator.
orden de una matriz cuadrada. order of a square matrix.
orden de un determinante. order of a determinant.
panescópico. panscopic.
parámetro convencional. conventional parameter.
partición de un conjunto. partition of a set.
percentil. centile; percentile.
pivote. pivot.
platicúrtico. platykurtic.
población. population.
polígono de influencia. polygon of influence.
polígono de proximidad. proximal polygon.
polígono de Thiessen. Thiessen polygon
polígono de Voronoi. Voronoi polygon
polinomio de grado *n*. polynomial of degree *n*.
polinomio en varias variables. polynomial in several variables.
polinomio ortogonal. orthogonal polynomial.
polinomios de Hermite. Hermite polynomials.
polinomios de Laguerre. Laguerre polynomials.
poliscópico. polyscopic.
precisión. precision.
predicción. prediction.
principio de la característica regional. regional characteristic principle; regional magnitude principle.
probabilidad. probability.
probabilidad condicionada. conditional probability.
probabilidad de transición. transition probability.
problema directo. direct problem; forward problem.
problema inverso. inverse problem.
problema normal. normal problem.
proceso aleatorio. random process.
proceso aleatorio factorizable. factorable random

process.
proceso aleatorio ortogonal. orthogonal random process.
proceso aleatorio simultáneamente ortogonal. jointly orthogonal random process.
proceso de Poisson. Poisson process.
proceso de Wiener. Wiener process.
proceso de Wiener-Levy. Wiener–Levy process.
proceso estocástico. stochastic process.
proceso markoviano. Markov process.
producto de un escalar por una matriz. product of a scalar and a matrix.
producto escalar. scalar product.
producto interior. inner product.
producto matricial. product of matrices.
promedio. average.
promedio aritmético. arithmetic average.
promedio geométrico. geometric average.
propiedad ergódica. ergodic property.
prueba. trial.
puntaje estándar. standard score.
puntaje z. z-score.
punto de acumulación. accumulation point.
realización. realization.
recta de regresión. regression line.
recursividad. recursion.
red. grid.
regresión. regression.
regularización. regularization.
relación de Cartier. Cartier's relation.
representación de una función aleatoria intrínseca. representation of an intrinsic random function.
representación móvil. sliding representation.
reservas in situ. reserves in place.
reservas recuperables. recoverable reserves.
residuo. residual.
residuo óptimo. optimal residual.
residuo reducido. reduced residual.
resistente. resistant.
restricción universal. universal condition.
resultado. outcome.
reticulado. grid.
rezago. lag.
robusto. robust.
rodograma. rodogram.
ruido blanco. white noise.
secuencia de Cauchy. Cauchy sequence.
selectivo. selective.
semivariograma. semivariogram; demivariogram.
semivariograma acumulado. graded semivariogram.
semivariograma compuesto. nested semivariogram.
semivariograma cruzado. cross-semivariogram.
semivariograma cúbico. cubic semivariogram.
semivariograma de de Wijs. de Wijsian semivariogram.
semivariograma de efecto agujero. hole effect semivariogram model.
semivariograma de efecto de pepita puro. pure nugget effect semivariogram model.
semivariograma de Gauss. Gaussian semivariogram.
semivariograma esférico. spherical semivariogram.
semivariograma experimental. experimental semivariogram.
semivariograma exponencial. exponential semivariogram.
semivariograma generalizado. generalized semivariogram.
semivariograma imbricado. nested semivariogram.
semivariograma lineal. linear semivariogram.
semivariograma logarítmico. logarithmic semivariogram.
semivariograma parabólico. parabolic semivariogram.
semivariograma penta-esférico. penta-spherical semivariogram.
semivariograma probabilístico regional. probabilistic regional semivariogram.
semivariograma promedio. average semivariogram; mean semivariogram.
semivariograma regional. regional semivariogram.
semivariograma regularizado. regularized semivariogram.
semivariograma relativo. relative semivariogram.
semivariograma relativo general. general relative semivariogram.
semivariograma relativo local. local relative semivariogram.
semivariograma relativo por parejas. pairwise relative semivariogram.
semivariograma subyacente. underlying semivariogram.
semivariograma teórico. theoretical semivariogram.
serie temporal. time series.
sesgo. bias.
sesgo condicionado. conditional bias.
sesgo nulo condicionado. conditional nonbias; conditional unbiasedness.
significado objetivo. objective meaning.
simulación. simulation.
simulación condicionada. conditional simulation.
simulación secuencial de un indicador. sequential indicator simulation.
síntesis de Fourier. Fourier synthesis.
sistema no singular. nonsingular system.
sistema regular. regular system.
sobreestimación. overestimation.
soporte. support.
soporte geométrico. geometric support.
spline. spline function.
suavizamiento. smoothing
subdivisión de Delaunay. Delaunay subdivision.
subestimación. underestimation.
suceso. event.

tamaño de la muestra. sample size.
tendencia. drift; trend.
teorema de Bayes. Bayes' theorem.
teorema de la aditividad. additivity theorem.
teorema de la proyección. projection theorem.
teorema del límite central. central limit theorem.
teorema de Pitágoras. Pythagorean theorem.
teorema ergódico. ergodic theorem.
teoría. theory.
teselado de Dirichlet. Dirichlet tesselation.
teselado de Voronoi. Voronoi tesselation
transformación. transformation.
transformación afín. affinity transformation.
transformación espacial de primera especie. space transformation of the first kind.
transformación espacial de segunda especie. space transformation of the second kind.
transformación normal. normal score transform.
transformación uniforme. uniform transformation.
transformada. transform.
transformada de Fourier. Fourier transform.
transformada de Laplace. Laplace transform.
transformada de orden secuencial. rank order transform.
transformada rápida de Fourier. fast Fourier transform.
transitivo. transitive.
translación de coeficientes autorizados. translation of permissible coefficients.
triangulación de Delaunay. Delaunay triangulation.
univariable. univariate.
universal. universal.
universo. universe.
validación. validation.
validación cruzada. cross-validation.
validación cruzada conjunta. joint cross-validation.
valor atípico. outlier.
valor errático. outlier.
valor esperado. expected value.
valor esperado condicionado. conditional expectation.
valor estimado. estimate.
valor promedio. average value.
valor propio. eigenvalue.
valor propio de una matriz. eigenvalue of a matrix.
variable aleatoria. aleatory variable; random variable; variate.
variable aleatoria continua. continuous random variable.
variable aleatoria de segundo orden. second order random variable.
variable aleatoria discreta. discrete random variable.
variable estocástica. stochastic variable.
variable regionalizada. regionalized variable.
varianza. variance.
varianza de dispersión. dispersion variance.
varianza de estimación. estimation variance.
varianza de extensión. extension variance.
varianza de krigeage. kriging variance.
varianza poblacional. a priori variance.
variedad. manifold.
variedad lineal. linear manifold.
variografía. variography.
variograma. variogram.
variograma cruzado. cross-variogram.
variograma de orden α. variogram of order α.
variograma de orden secuencial. rank order variogram.
vector. vector.
vector ortogonal. orthogonal vector.
vector ortonormal. orthonormal vector.
vector propio. eigenvector.
ventana. window.
zona de influencia. zone of influence.

SAM HOUSTON STATE UNIVERSITY
0 2118 00754 1244

REFERENCE

WITHDRAWN

Building Use Only

D41000625553